Reading through the Charcoal Industry in Ethiopia:

Production, Marketing, Consumption and Impact

Melaku Bekele and Zenebe Girmay

FSS Monograph No. 9

Forum for Social Studies (FSS)
Addis Ababa

Printed in Addis Ababa

Typesetting & Layouts: Konjit Belete

ISBN 13: 978-99944-50-48-0

Forum for Social Studies (FSS)
P.O. Box 25864 code 1000
Addis Ababa, Ethiopia
Email: fss@ethionet.et
Web: www.fssethiopia.org.et

This Monograph has been published with the financial support of the Civil Societies support program (CSSP). The contents of the Monograph 9 are the sole responsibilities of the author and can under no circumstances be regarded as reflecting the position of the CSSP or the FSS.

Table of Contents

Acknowledgement

We shall extend our gratitude to Forum for Social Studies (FSS) for giving us the opportunity to study the charcoal issue in Ethiopia. We also acknowledge the overall support by the Executive Director, and the administrative staff of FSS during our work. Particular appreciation shall go to FSS research staff for constructive input provided to further enhance the quality of this manuscript. We thank all those who provided us with information, help in data collection and facilitate field visits.

List of Tables

List of Figures

Abbreviations and Acronyms

ARDB	Agricultural and Rural Development Bureau
AREAP	African Renewable Energy Access Program
ARPADB	Afar Regional Pastoral and Agriculture Development Bureau
BoFED	Bureau of Finance and Economic Development
Cal	Calorie
CRGE	Climate – Resilient Green Economy
CSA	Central Statistics Agency
EFAP	Ethiopian Forestry Action Program
EPA	Environmental Protection Authority
EREDPC	Ethiopian Rural Energy Development and Promotion Center
FAO	Food and Agricultural Organization
FDRE	Federal Democratic Republic of Ethiopia
GDP	Gross Domestic Program
GHG	Green House Gases
GTZ	Deutsche Gesellschaft für Technische Zusammenarbeit (German Technical Cooperation)
IAP	Indoor Air Pollution
LPG	Liquefied Petroleum Gas
MDG	Millennium Development Goal
MJ	Mega Joule
MoA	Ministry of Agriculture
SNNPRS	Southern Nations, Nationalities and Peoples Regional State
SSA	Sub-Saharan Africa
WBISPP	Woody Biomass Inventory and Strategic Planning Project
WHO	World Health Organization

Executive Summary

Ethiopian cities and towns burn over three million tons of charcoal each year. Charcoal is 99% flammable when dry, cheaper compared to modern sources, and also accessible. Charcoal is easy to transport, efficient and produces a steady heat with little or no smoke. In Ethiopia, it is the poor who are engaged in charcoal making and distribution. Dependency on charcoal is rather increasing as a result of rapid growth in urban population, and rise in price of modern sources of energy like electricity, LPG and kerosene. It is a source of cash income for rural households with little or no land rather than source of energy; many urban youth, and women in particular are engaged in the retail business.

Studies in many African countries show that charcoal making is among the primary drivers of deforestation and subsequent land degradation. In the case of Ethiopia, charcoal is produced from state-owned (public) forests and woodlands. There is little regulatory intervention from the government side. Moreover, production is more traditional and the producers have little idea that charcoal can be produced efficiently with modern technologies. Although charcoal meets significant portion of urban households' energy needs in the country, and also support the livelihood of tens of thousands of rural households, it hardly attracted the attention of policy makers and development agents. A good majority of urban population who use charcoal on regular basis doesn't seem to know how charcoal is made, from where it comes, and its adverse environmental impacts.

In cognizant of the potential environmental impact of charcoal production and marketing in the country, FSS commissioned this study with the objective to understand the environmental, social and economic implications of charcoal production, marketing and consumption in Ethiopia with aim to generate/increase awareness among the general public and incite a policy debate among concerned key stakeholders.

The areas that were included in this study cover most of those places known for their large scale charcoal production in the country, Gewane in Afar, Bilate in SNNPRS, and Langano in Oromiya. Addis Ababa and the major regional cities/towns are considered in the assessment as the major charcoal consumption centers.

This study used both primary and secondary sources. Collection of primary data involved extensive fieldwork in different regions between July and November 2012 in which such tools like questionnaires survey, key informant interviews, field observations, and on-site demonstrations are employed. The field assessment covered producers, distributors, wholesalers, retailers, and consumers, governmental and non-governmental bodies. The secondary sources include travelers' accounts, government documents and research reports.

The study result shows that charcoal production, transportation, and distribution remain a risky and highly inefficient undertaking. The current charcoal business in the country is unsustainable and has dismal pictures. It has failed to attract solid investments that are necessary for research and development; the banning of the making and transportation of charcoal remained ineffective in regulating the production and trade. As a result, thousands of small-scale producers, transporters and distributors cling to the business out of sheer need for survival. With the exception of some women engagement in the retailing activity, the charcoal business in Ethiopia appears to be dominated by illiterate young men.

The dominant charcoal production technology remained the traditional kiln. Despite their promising results in improving the quality and production efficiency of charcoal, the adoption of some improved charcoal making technologies, e.g., Casamance kiln, metal kiln, drum kiln, is yet confined to certain areas and has limited applicability owing to the substantial investment requirements. On the other hand, the production of charcoal briquette from different agricultural and/or forest wastes has started by the Ethiopian Rural Energy Development and Promotion Center (EREDPC) on a limited scale. Biomass wastes from bamboo, *Prosopis juliflora,* cotton stalk, *Chat* (*Khat*) stem and coffee husks have shown promising results. Given its great potential for converting waste biomass into fuel for household use, in an affordable and environmentally friendly manner, the charcoal briquetting using agricultural wastes and other materials can be a viable alternative to wood charcoal.

The bulk of charcoal entering Ethiopian towns and cities is produced in the acacia-dominated dry-woodlands of the country, which have been over-exploited freely for decades as the property rights on these resources are loosely established and/or there is little control over the resource base. In Ethiopia, charcoal production heavily depends on acacia species for the quality they constitute of. A considerable amount of charcoal is also being produced from the invasive species–*Prosopis juliflora* in Afar Regional State, which supplied to nearby towns, mainly to the capital Addis Ababa.

With mounting urbanization, population growth and economic development on one hand, and the absence of affordable and convenient modern alternative energy sources on the other, the switch from firewood to charcoal will continue at higher rates. Besides its convenience and accessibility at reasonable cost as household energy source, charcoal trade is also offering important income generation opportunity. Hence, charcoal will expectedly remain the main cooking fuel for most people in the country's towns and cities for the foreseeable future. The assessment also showed that the overall trend in production, consumption and price of charcoal is found to be increasing in the regions' major towns and cities, particularly in Addis Ababa, the largest consumption and marketing center. Reports showed that the volume of charcoal produced in Ethiopia increased to about 3.6 million tons in the year 2009 from an estimated amount of 3,320,535 tons of charcoal between 1995 and 2005. A charcoal inflow survey (conducted in August 2012) to the city of Addis Ababa alone showed that an estimated 42,045 sacks of charcoal, suggesting an equivalent of 537,124.875 tons of charcoal per annum, has entered the city in a day. Most of the charcoal (about 55%) entering Addis Ababa comes through the eastern gate (Kaliti), i.e., Awash, Gewanie and Bure-Dimtu in Afar, by ISUZU medium duty trucks. In Ethiopia, the current charcoal production system does not take the tree resource into account; charcoal is illegally produced from free sources. Thus, the main actors along the channels of charcoal supply to urban consumers in the city of Addis Ababa are producers, distributors/transporters, wholesalers, and retailers.

In a similar vein, the broad assessments made in the major regional towns showed similar situations. The charcoal business in the various regions is dominated by illegal actors (including some illegal export trade across borders as in Afar and Somali region), and traditional production technologies (mostly by the landless youth), with no incentives for investment in planned and sustainable ways and weak organization and control; while the production, demand, and consumption is increasing at the expense of the dwindling forest resources.

Even though there is lack of reliable information, fuelwood extraction (firewood and charcoal) is often associated with the alarming rate of deforestation and environmental degradation in Ethiopia. The prevailing charcoal production systems in Ethiopia are unsustainable; the raw materials for charcoal come from free sources (with reckless cuttings and little culture of plantations), and the production technology (which uses the *traditional charcoal kilns with an efficiency ranging only between 10 and 15%*) is highly inefficient. Thus, the massive wastage of standing wood stocks has a direct link to the worsening of forest depletion (most of which are poorly managed and prone to degradation), soil degradation and environmental degradation, which in turn deteriorates the quality and quantity of various ecosystem services at large.

Besides depleting the forest resources that would have sequestered carbon into their body mass, charcoal production phase (of the inefficient traditional technologies) is known to emit various GHGs (e.g. CO, CO_2, CH_4), hence contributing to climate change. Charcoal has also considerable health impact to producers, during the carbonization process, and to consumers upon indoor combustion through emitting smoke and various gaseous mixtures (mainly CO).

The study concluded that the major shortfall in the charcoal industry in Ethiopia is the institutional deficits it has been suffering from for a long time. There is no public agency or any kind of regulatory intervention on the part of the government to regulate the production, marketing, consumption, as well as impact of the charcoal industry in the country. Charcoal is produced and marketed in a policy as well as legal vacuum. This is the apparent failure of the concerned public agency or agencies not only over a charcoal issue, but the policy collapse of forest and woodland management in the country. The most familiar intervention on the part of public agencies is the criminalization of charcoal producers with little result to stop them.

Charcoal is one of the many forest products. As millions of people remained dependent on it as sources of energy and income the state's institutional intervention becomes critical and overdue. Therefore, the negative illustration of charcoal production as a cause for environmental degradation should change and looked at as one of forest products that need to be regulated.

To improve the condition in the charcoal industry, first the issue should be set as a policy agenda to dialogue over: the potential of the industry to create job opportunities must be recognized; the need to end the open access situation of the woodlands and putting a property arrangement over the resources; the introduction of a management system in which exploitation can be based on the capacity of the resource to recover itself; giving charcoal its own source by establishing forest plantations of appropriate species; creating a charcoal agency to regulate the industry, work towards improving the charcoal technology and diversify its sources; de-criminalize charcoal production and include charcoal in the extension packages. The ultimate push must include the development of modern energy sources. Education and research should also focus on improving efficiency in production and marketing of charcoal.

In conclusion, what the charcoal industry requires most is an institutional recognition on the part of the government as a viable sector to create jobs, and serve millions of people as source of energy and income. Then the rest follows.

1. Introduction

1.1. Background

Nearly half of the world's population and about 81% of Sub-Saharan African (SSA) households rely on wood-based biomass energy (firewood and charcoal in particular) for cooking and heating (AREAP, 2011). Wood-based biomass as the main source of energy is reported at 68% in Kenya, 95% in Eritrea, 94% in Ethiopia, while 70% and 92% is indicated for Zambia and Uganda, respectively (van Beukering, 2007). Fire-wood and charcoal accounted for about 91% of Africa's round wood[1] production in 2000 (Falcão, 2008).

Charcoal, which is scarcely used in the rural areas because of accessibility of "free" wood, is quite popular in urban centers because of ease to use compared to firewood (FAO, 1993; Luoga et al., 2000). According to Madon (2000), urban women interviewed during household energy surveys in Ethiopia, Chad, Madagascar, Mali, the Niger and Senegal did not like to cook with wood because they found it difficult to kindle, awkward, dangerous for children, smoky and messy. Charcoal is perceived to lack most of these negative effects, and it is priced less than liquefied petroleum gas[2] (LPG) and kerosene, which are still too expensive for many people (Foster, 2000). Rapid urbanization, increasing poverty and high population growth rates are driving the growth in the use of charcoal in urban cities and peri-urban areas (Girard, 2002).

The charcoal business[3] employs a large portion of the population along the chain from the producer in rural areas to the distributors and retailers in urban areas. In

---

[1]Round wood production (in forestry) comprises all quantities of wood (a length of cut tree often with round cross-section such as logs, poles etc) removed from the forest and other woodland, or other tree felling site during a defined period of time for industrial or consumer use.

[2]When it comes to charcoal, domestic energy preference does not always follow price fluctuation. A study by Ibrahim (2003) in Sudan showed that households prefer charcoal for its unique cooking properties when even price is three times higher compared to such energy sources like LPG.

[3]Charcoal is alleged to finance organizations like Al-Shabaab in Somalia. According to the Christian Science Monitor (September 21, 2012), a UN agreement to buy charcoal for cooking food for African Union troops may indirectly be funding Al-Shabaab. Moreover, the paper quoted a senior US State Department official: "when charcoal flows out to Yemen, to Saudi Arabia, to places in the Gulf, Al-Shabaab is able to tax this charcoal and gain the resources from it."

1

some cases, 60-80 percent of rural household income is generated from charcoal making and trade (Chidumayo and Emmanuel, 2010). The contribution of biomass fuels in national energy utilization in the majority of dry forest and woodland countries is huge, ranging between 35-75 percent in many countries (e.g. Senegal, Togo, Ivory Coast and Angola) and over 75 percent in many others (e.g. Sudan, Ethiopia, Eritrea, Kenya, Tanzania, Mozambique, Zambia, DRC and Nigeria) (van Beukering, 2007; Malimbwi et al., 2010).

Studies indicate that dependency on wood-based fuel will continue to be the dominant source of energy for developing countries, particularly the SSA-far more than in any other region in the world (AREAP, 2011). Tomaselli (2007) estimated a growth of 3.7% annual charcoal consumption in SSA. These trends coupled with inefficient charcoal production and consumption practices, and inaccessibility by most households to modern energy types signify the continued and probably growing dependence on the already dwindling biomass resource for energy (Kwaschik, 2008).

If there is no alternative option to biomass wood consumption for fuel, the number of people dependent on biomass will increase to over 2.6 billion by 2015 and to 2.7 billion by 2030 due to population growth (OECD/IEA, 2006). Access to electricity and LPG is not expected to replace wood-based fuel for cooking in the near future as the cost of using these energy sources are often too expensive (van Beukering et al., 2007). Nevertheless, the United Nations Millennium Project has made it its goal to reduce the number of households using traditional biomass for cooking by half by 2015. This will involve 1.3 billion people switching to other fuels (MDG) (OECD/IEA, 2006).

In Ethiopia, charcoal is an indispensable renewable[4] energy source. It is relatively cheaper and accessible. Biomass fuel (in the form of firewood and charcoal), which supplies about 90% of Ethiopia's energy needs, is the biggest source of energy in the country (Yisehak and Duraisamy, 2008). Charcoal production and consumption provides valuable employment and income, mainly to vulnerable groups in society. It is an extremely important economic safety–net

---

[4]Renewable energy refers to the natural energy sources that are always available to be tapped (were not formed and never run-out), which includes water, wind, sun and biomass (vegetation). Biomass resources include trees, food crops, agricultural and forestry by-products; and since charcoal is biomass by-product, it is treated under renewable energy sources.

as it provides readily available cash for food and livelihood security for the very poor in rural areas (Daniel, 2005).

As important as it is for employment and energy source, charcoal is also causing serious damage to the environment. As early as 1984, the World Bank (1984) indicated that fuel-wood gathering is one of the most important causes of deforestation, resulting in the clearance of about 10 Million hectares of forest each year in the developing world. Most of extraction for fuel-wood is made from naturally occurring forests free of charge. This led to the decline of forest cover and fuel-wood scarcity in many countries (Lopez, 1997). Due to shortage in fuel-wood, countries like Ethiopia, are more and more using crop residues and animal dung as fuel, a practice that reduces the availability of valuable nutrients for the soil (Hawando, 1997).

Increased use of agricultural residues and animal dung deprives the land of essential nutrients that are necessary for soil fertility (World Bank, 1984; Hawando, 1997). This means, agriculture, which is the mainstay of the economy, is negatively affected by the existing energy consumption pattern. It is estimated that nutrient loss and soil erosion result in the loss of close to 600,000 tons of grain per year and this is equivalent to 90 percent of Ethiopia's food deficit in 1993 (World Bank, 1984). In an earlier study, the World Bank (1984) estimated that the growing diversion of natural fertilizers in dung and crop residues to household fireplaces reduced crop yields by more than one million tons of grain a year in Ethiopia. The loss of soil fertility and land degradation leads to financial loss of about 2% of GDP in Ethiopia, according to EFAP (1994).

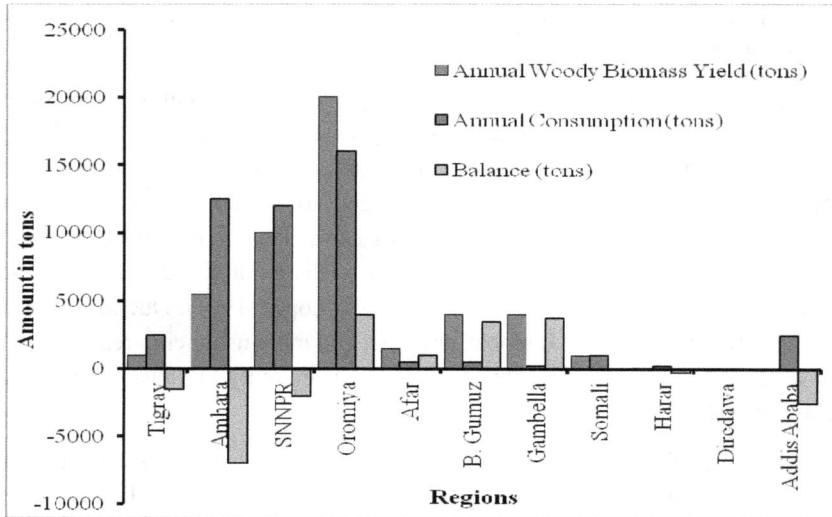

Figure 1: Fuel-wood situation in Regional States of Ethiopia

SOURCE: Kiflu et al., 2009

As illustrated in Figure 1, there is a negative balance between biomass energy consumption and supply in most part of the country. As a result, the cost of fuel-wood increase is challenging the already staggering living condition.

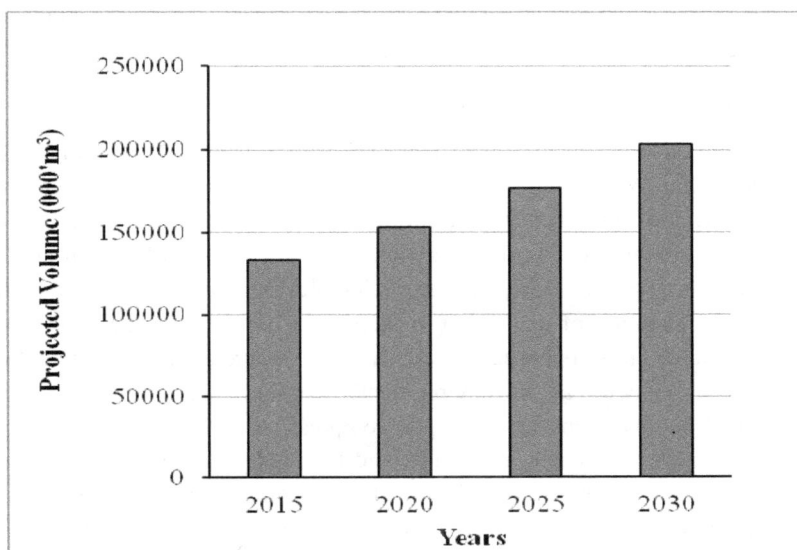

Figure 2: Future demand for forest wood products (Ethiopia)

SOURCE: FOSA, 2000

In Ethiopia, as demand for charcoal increases with rapid urbanization (Figure 2), so does pressure on forests and woodlands, most of which are poorly managed and prone to degradation. The current practices in the industry related to charcoal production, trade and consumption are unsustainable. The raw materials for charcoal come from free sources (without charge), the production technology is inefficient with conversion ratio between 10 and 15% (Yisehak and Duraisamy, 2008), the trade is unregulated, and the impact on the environment and human health seems huge. The dry woodland ecosystems, where most preferred acacia tree species for charcoal grow, constitute huge potential for economic development. But, they are caught in a spiral of deforestation, fragmentation, degradation and desertification due to, mainly, various human-induced causes (FAO, 2010).

The observation that the charcoal industry does not receive the policy attention it deserves, and, as a result, suffers from structural governance deficits may in part be explained by lack of inclusive information concerning the industry's livelihood significance and source of energy for millions of people. Hence, investigating and understanding the industry's significance to society at large

5

and identifying its limitations will help to create awareness among the public and initiate policy dialogue on the part of the government.

1.2. Objective and Scope of the Study

Charcoal production in developing countries in particular is increasing rather than decreasing. According to FAOSTAT (2011), Ethiopia is one of the three countries (next to Brazil and Nigeria) of the world known for their high charcoal production. Although charcoal meets significant portion of urban households' energy needs and generates large number of employment in the country, it hardly attracted the attention of any policy makers for a long time. The majority of urban consumers, it is supposed, are also unaware where the charcoal they burn comes from, as well as its environmental consequences. Hence, charcoal dependent Ethiopia needs to develop a compressive policy and invest in the industry to make it efficient, profitable, and less damaging to the environment and human health, and ultimately build a sustainable business that could generate employment and source of energy in an environmentally friendly manner. Policies like that of charcoal need to be constructed on tangible data for successful implementation of programs.

It is, thus, the general objective of this assessment to generate relevant information concerning the environmental, social and economic implications of the charcoal industry. The study focuses on the production, marketing and consumption patterns of charcoal in Ethiopia, with the aim to contribute towards improved understanding and increased awareness, and incite policy dialogue among the general public, government agencies, and other relevant bodies.

1.3. Study Areas and Methods

1.3.1. Study Areas

To get a comprehensive picture of the charcoal industry in Ethiopia, some of the major charcoal production, marketing and consumption centers were considered during the field surveys (Figure 3). Addis Ababa was put at the center of this study with the assumption that it is the biggest center for charcoal marketing and consumption in the country. Field surveys, including personal stories of producers and on-site demonstrations of charcoal making processes, were conducted in the major charcoal producing areas; namely Gewanie (Afar), Langano (Oromiya) and Bilatie (SNNPR). These charcoal production centers are reported to be the main charcoal suppliers to nearby towns and Addis Ababa, which is the major consumer.

Broad assessments were also undertaken in major regional towns regarding charcoal issues. These include; Mekelle (Tigray Regional State), Bahirdar (Amhara Regional State), Awash and Gewanie (Afar Regional State), Adama (Oromiya Regional State), Harshin/Jijiga (Somali Regional State), Hawassa and Arba-Minch (SNNPR), and Diredawa city administration.

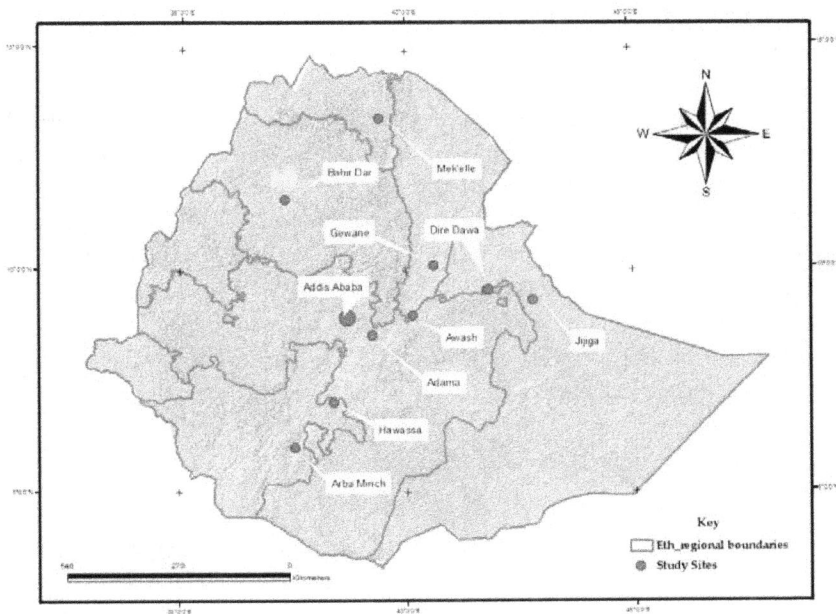

Figure 3: Location of the study areas

1.3.2. Methods

The monograph was compiled using both primary and secondary sources. To collect the primary data, extensive field surveys were conducted between July and November 2012 in the major charcoal production, marketing and consumption areas. Primary data collection employed various tools, including questionnaires, interviews, field observations, on-site demonstrations and individual experiences.

7

In the field survey, producers, distributors, wholesalers, retailers, consumers, governmental and non-governmental bodies were included. Correspondingly, the assessment considered various aspects of the charcoal industry: production, supply, demand, consumption, marketing, impact and other related issues. To address all these important charcoal issues in the country, the field assessments were conducted in three separate phases.

Phase-I

The first phase of the fieldwork was conducted in the capital Addis Ababa–the, supposedly, biggest charcoal marketing and consumption centre in the country. A relatively detailed field surveys were undertaken in the city of Addis Ababa to assess the supply, demand, consumption, marketing and other related issues of charcoal. Prior to the actual data collection, a reconnaissance survey was conducted to have an overview on the charcoal industry in the country in general, and the city of Addis Ababa in particular. For this purpose, some concerned bodies, including the Environmental Protection Authority (EPA), Agricultural and Natural Resources Management Offices (Forestry departments), Forestry Research Centre (FRC), the Ministry of Water and Energy were consulted. However, there was found no single responsible body which regulates the charcoal industry; the sector is not legalized either. It was, thus, difficult to quantify and locate the target groups (actors) for this research: the transporters/distributors, wholesalers (depot owners) and retailers.

As a result, the researchers decided to cover all the sub-cities (kifle-ketemas) of the city employing quota sampling technique to select respondents. Target groups of great interest to the study were first identified and then sample respondents were selected. These include; the distributors, wholesalers, and retailers. A pre-testing survey was carried in one kifle-ketema to observe the distribution of wholesalers and retailers. Then, the number of respondents for each of the wholesalers and retailers was arbitrarily set at 100 (assuming 10 respondents for each stratum from each kifle-ketema). However, the number of distributors was not fixed since they enter into the city during night time and are not readily accessible; all possible effort was used to cover as much distributors as possible.

Even though the study was originally envisaged to include only the main actors in the charcoal business–the distributors, wholesalers and retailers, it was later recognized that the involvement of some consumer households is also important to corroborate issues on consumption and demand of charcoal. As a result, four households from each sub-city (a total of about 40 households) were designed to be included.

To facilitate data collection, three field assistants, who have experience on similar data collection, were identified and trained. Afterwards, a separate pre-tested checklist of questions were distributed for each target group to gather data on the supply, demand, consumption, marketing and other related issues of the charcoal industry in the city. Finally, the required data was collected from a total of 239 respondents; 100 wholesalers, 95 retailers, 7 distributors and 37 consumers.

A separate charcoal inflow survey to Addis Ababa was also conducted for the period between August and September, 2012. It was designed to know the inflow of charcoal to Addis Ababa through its major inlets. A 24 hours follow up and recording of sacks of charcoal, supplied using various types of trucks, was conducted for four consecutive days in each of the five major inlets (Kaliti, Sebeta, Sululta, Burayu, and Kara) in August 2012.

Phase-II

In the second phase, charcoal production and related issues were assessed in the main charcoal producing centers. As pointed out during interviews with respondents in Phase-I, it was recognized that majority of the charcoal consumed in the city of Addis Ababa comes from nearby districts of Oromiya Regional State (mainly Langano), Afar Regional State (mainly Awash and Gewanie) and sometimes from SNNPR Regional State (mainly Bilatie district). Consequently, interviews with producers, including on-site demonstrations on charcoal production processes and the associated impacts, were made in these major charcoal production centers (Awash, Gewanie, Langano and Bilatie).

Phase-III

Lastly, relatively, broad assessments were extended into some of the major regional towns (Mekelle, Bahirdar, Awash, Gewanie, Adama, Jigjiga, Hawassa, Arbaminch, and Diredawa city administration) to complete the picture of the charcoal industry in the country. A set of similar checklist of questions were distributed into these areas, with close follow-ups by researchers and assigned facilitators in each site.

The researchers have visited all the identified major production centers including Awash, Gewanie, Langano and Bilatie where they observed the production processes, and possible impacts, and interviewed producers on various aspects of the charcoal making. They also visited some of the regional towns (Mekelle, Hawassa, Arba Minch), while others were addressed through assigned facilitators. For Jigjiga, as data collection through assigned facilitator failed, charcoal issues in the region were displayed mainly using the available literatures.

Besides, information on various issues of charcoal was collected from the available literatures. For this purpose, various secondary sources, including journals, books, proceedings, government reports, travelers accounts and others were used.

The quantitative data was organized and fed to Microsoft excel and subjected to descriptive statistics. Results were interpreted and displayed in percentages, graphic and tabular illustrations.

1.4. Structure and Limitation of the Study

The study starts with general information about the world/Africa and Ethiopia's fuel-wood production. This is followed with a brief historical review of fire-wood and charcoal production, consumption and the long standing scarcity in Ethiopia by referring to travelers account and a few available early research findings. The subsequent two sections deals with production, marketing and consumption of charcoal in which charcoal making process, the chemistry and quality of charcoal, its sources, the producers' profile, the market chain and consumption patterns are presented and discussed. In the last two sections, impact of charcoal on the environment and human health are presented, and recommendations based on findings are put forward.

The study has a limitation in providing detailed and comprehensive pictures about profit distribution in the market chain in Ethiopia. Impacts of charcoal on human health and the environment are based mainly on limited literature available. Due to the lack of attention to the charcoal industry, there is no government body at the Federal level that could provide us with information about the charcoal industry (production, trade, and marketing) in the country as well as any future plan about the sector.

2. Fuel-wood Production and Consumption in Ethiopia: Historical Review

2.1. Fire-wood

In Ethiopia, just like many other African countries, fire-wood was collected where it was free and available with little restriction. Branches, dead-wood and leaves were often used for domestic consumption, while most freshly cut and prepared wood ended in the markets of urban centers. However, these days, branches, dead-wood and leaves are also sold in local markets.

During the Imperial period, beside state owned forests, part of fire-wood sold in towns was commonly collected from privately owned natural forests whose owners gave permissions of utilization to their tenants in return for a certain load of wood to be carried to their residences. As most of the owners were absentee landlords, and as the forests were far from any kind of permanent guard, it was not always necessary for tenants to secure the landlords' permission. At the time, it was in the interest of the land owners who usually wanted to see their land cleared of forest trees to start food crop cultivation (Melaku, 1992).

Figure 4: Wood collector, Ethiopia

SOURCE: www.das-fotoarchiv.com

FAO provides some figures for round wood and fuel-wood production in Ethiopia for the years from 1949 and 1973 with great variation through the years. For instance, round wood production was recorded at 8.1 million m3, and from 1950 to 1957 it remained at 8 million m3. In 1960, the estimate suddenly rose to 29.4 million m3 and dropped to 20 million m3 in 1963 (see Table 1).

Table 1: Total round wood and fuel-wood production in Ethiopia for some selected years (in 103 m3)

Year	Total round wood	Fire-wood and charcoal	% of total
1949	9,200	8,120	88.2
1953	8,062	8,000	99.2
1960	29,452	29,400	99.8
1963	20,470	19,500	95.2
1965	21,006	20,000	95.2
1967	21,537	20,500	95.1
1969	22,575	21,500	95.2
1970	23,105	22,000	95.2
1973	24,220	23,000	95.0

SOURCE: FAO Year Book of forest products, 1950 -1974

Although the above data may not be reliable, it is the only available information and provides a general picture of wood production for fuel in the past. What these figures indicate is that the bulk of wood is employed for fuel-wood. Out of the total round wood produced between 1949 and 1973, about 96 % was used for fuel-wood (fire-wood and charcoal). The industrial share from the total wood produced in 25 years was limited to only 3.2 % of the total.

Other sources referring to fuel-wood consumption of Ethiopia include that of Logan (1946) and Russ (1944). The former estimated 10 million cords (about 35.5 million m3) as the total Ethiopian fuel-wood consumption taking into consideration that part of the population who were using cow-dung in the 1940's. The American forester, David Russ (1944), on the other hand, estimated the total annual fuel-wood consumption at only 4 million m3, which is much lower than the figures provided by Logan (1946). These inconsistent and conflicting figures on the country's fuel-wood supply contribute to the difficulty of making effective policies and programs.

Table 2: Wood-fuel consumption estimate for 1983 (Ethiopia)

Purpose	Volume (million m3)	Percent
Firewood	16.70	79.60
Charcoal	2.10	10.0
Construction	1.00	4.80
Poles	1.00	4.80
Industrial use	0.17	0.80
Total	20.17	100.0

SOURCE: (Newcombe, 1983)

On the basis of 18 million population estimate of Ethiopia for 1950 (Pankhurst, 1961) and on the assumption that per capita fuel-wood consumption was about 0.57 m3 per year (taking the 1983 per capital consumption of fuel-wood estimate by MoA (1986)), a total consumption of 9 million m3 of wood may be closer to the reality for the country in the 1950s (Melaku, 1992). As there were few managed forests, fire-wood was extracted from near open-access state-owned and private natural forests, which results in the depletion of the resource base. Sustainable[5] round wood production can only be ensured under well managed forests.

2.2. Charcoal

There is little evidence to present if charcoal had been in use in ancient towns of Ethiopia. It is, however, evident that the long tradition of ironwork must have required the production and use of charcoal in the country. A relatively large scale charcoal use must have started, however, with the emergence of new settlements or towns, particularly during and after the last quarter of the 19th century.

An official document about charcoal consumption (1927-1933) for Addis Ababa (Table 3) indicates that 6,093 tons of charcoal was brought to Addis Ababa

---

[5] In sustained forestry, the annual timber production of nearly 21 million m^3, according to Pohjonen (1988), might correspond to 5 million ha of well managed, indigenous forests (with mean annual increment of 44m^3/ha/y or one million ha of fast growing eucalyptus plantations at the rate of 20m^3/ha/y mean annual increment.

between 1927 and 1933 from the acacia woodlands in Modjo, Adama and Wolenchiti by individual concessionaires (Mooney, 1954; MoA, 1986). (Terms of concessions were not mentioned).

Table 3: Charcoal brought to Addis Ababa from 1927 to 1933

Year	Quantity of charcoal	
	Bag of 45 kg	In tons
1927	11,247	506
1928	15,483	696
1929	19,423	874
1930	14,453	650
1931	25,635	1,153
1932	28,515	1,283
1933	20,701	931

SOURCE: MoA, 1986

The official estimate of annual charcoal consumption for the country in the 1970s was 150,000 tons produced from 1.7 million m3 of wood. According to a 1983 estimate of wood removal for charcoal, annual consumption grew to about 190,000 tons with an average increase of 5,000 tons of charcoal per year between 1975 and 1983. Estimates were made on the basis of the amount of charcoal brought to bigger towns passing through check points, and did not take into account a substantial amount transported illegally to the capital. Acacia (Girar in Amharic) is the preferred wood for charcoal making at the time. High–density woods like acacia are preferred to produce charcoal of higher density and lower fragility, which burns more slowly at a higher temperature than charcoal from low-density wood (Uhart, 1975). Charcoal was also produced from Eucalyptus, Erica, Juniper and other trees.

During the Imperial period, there was relatively organized and large-scale charcoal production and distribution in the country. Five groups of people were involved in this process: the forest owner, charcoal producer, the transporter, distributor and the retailer. The transaction between the forest owner and the producer was valid only if a representative from the forestry department issued a tree utilization permit after demarcating the forest area and estimating the volume of wood to be felled (MoA, 1986). However, as operations in the forest were rarely supervised by the Forestry Department, tree felling regulations or the rules which required the owner of the lease to plant a number of seedlings in place of felled trees, were not usually observed (Melaku, 1992). Since the nationalization of all forests in 1975, charcoal making was declared illegal and all activities went underground. As the "forest owner" group in the market chain disappeared and charcoal started produced from free sources, the true market price of charcoal could not be re-established.

2.3. Fuel-wood Scarcity: Brief Account

Fuel-wood scarcity in various parts of Ethiopia has been recorded by several early European travelers. Some of the descriptions showed the gravity of the shortage in some part of the country already in the 16th century. In fact, in the northern and in some parts of central Ethiopia, where most of the original forest has long been lost and the culture of tree planting was very limited, many households have been using cow dung and crop residues for heating and cooking (at least since the 16th or the 17th centuries) (Beckingham and Huntingford, 1961). Bruce (1813) observed shortage of fuel-wood in Gondar in the early 1770s where people burn cow and mule dung as a substitute to wood. He stated: "only royalty and rich people could afford it". According to Harris (1844), "fuel-wood was carried by marching soldiers for later use during campaign". Vivian (1901), while traveling towards Addis Ababa from the northeast, claimed himself experiencing serious shortage of fuel-wood. He wrote: "For two or three days before reaching the capital we had to do without wood in camp, for there was scarcely a tree to be seen. Every shrub that could possibly be used for firing had been cleared off years ago". Wylde (1901) also noted that even though he found ducks on most ponds in northern Shewa, he did not kill them because of difficulty in cooking anything owing to want of fuel-wood. The French traveler (Merab, 1922) also noted about fuel-wood scarcity in Addis Ababa. Due to the scarcity, he said, cow-dung or even donkey droppings were used as fuel by poor people. According to the British forester, Logan (1946), fuel-wood scarcity attracted many people to the trade and there were, in addition to the normal peasant traders, established fuel-wood contractors who regularly supplied the local markets on a cash basis. Relatively recent fuel-wood scarcity assessments also provide comparable picture. For example, Wilson (1977) noted that the scarcity was such that, in northern Ethiopia, people dug out roots for fuel, a process which took a whole day to provide a donkey load of about 50 kg of fuel-wood.

3. Charcoal production

3.1. What is Charcoal?

Charcoal is a dark grey solid carbon residue obtained by removing water and other volatile constituents from vegetation substances created through the process called pyrolysis[6]/carbonization; the burning of carbonaceous raw materials in the absence of oxygen (FAO, 1985; AREAP, 2011). Without oxygen, the wood substance will decompose into a variety of substances, the main one of which is charcoal–a black porous solid consisting mainly of carbon (BTG, 2010; Brewer et al., 2010). In this process of incomplete combustion of plant materials, what is produced is not only charcoal, but also complex range of solid, liquid and gaseous products. Charcoal[7], however, is the most important product obtained following the pyrolysis of biomass.

Charcoal retains morphology of original feedstock, and burns without flame. Charcoal produced from hardwood, for example, is heavy and strong, whereas produced from softwood is soft and light (GIZ, 2012). Although wood is the most commonly used raw material, various types of biomass can be used to produce charcoal. These include: agricultural residues, sawdust, fruit stones, bark, cotton seeds, coffee husks, and wood shavings and sawmill residues and other similar products (FAO, 2008; Practical Action, nd; Nketiah, 2008).

3.2. Charcoal Quality

The quality of charcoal depends on both biomass used as raw material and the carbonization technology employed; and is defined by its physical, chemical and combustion properties. The desirable properties of quality charcoal as presented by FAO (1985) are: lower moisture content (between 5 and 10%), slow burning with higher calorific value (from 27 to 33 MJ/kg), higher fixed carbon content (from as low as 50% to as high as 95%), lower ash content (between 0.5 and 5%), and producing little smoke without objectionable nor toxic fumes and

---

[6]Pyrolysis/carbonization can be defined as the thermal decomposition of complex carbonaceous substances such as wood or agricultural residues in an oxygen deficient environment (FAO, 1987; BTG, 2010).

[7]Charcoal is not just pure carbon or a single compound, but composed of various elements: carbon (C), hydrogen (H), oxygen (O), nitrogen (N), sulfur (S), and ash, among others (Czernik, n.d). With the exception of charcoal, all of these materials are emitted with the kiln exhaust (Brewer et al., 2010; Czernik, nd).

neither spits nor sparks. These qualities are found in many Acacia species and some other woody species. For instance, Acacia amplecips, A. negrii and A. asak were reported to be the most preferred acacia species to produce quality charcoal (El-Juhany et al., 2001).

Generally, all woody species can be carbonized to produce charcoal; hence charcoal makers use a variety of tree species. As shown in Table 4 below, the quality of charcoal, however, varies from species to species, being dependent on the method of carbonization (Mugo and Ong, 2006).

Table 4: Approximate composition of some different charcoal types

Source of Charcoal	Moisture (%)	Volatile matter (%)	Ash content (%)	Fixed carbon (%)	Calorific value (cal/gm)
Acacia sp.	3.67	22.90	3.64	69.79	7780
Bamboo	9.31	15.03	14.80	60.86	6959
Prosopis	3.90	25.90	3.50	66.80	6256
Cotton stalk briquette	4.10	17.20	20.30	58.40	4588
Khat stalk briquette	8.04	28.58	16.54	46.84	5100

SOURCE: Yisehak and Duraisamy, 2008

According to producers visited in production sites in Gewanie and Bilatie, and several other studies (e.g. Damascene, 2005; Mugo and Ong, 2006; Falcão, 2008; Tinsae et al., 2012), the desirable qualities for fuel-wood species can be described as dense wood with low moisture content, relatively easy to cut and handle constituting of the qualities when carbonized and turned into charcoal. These criteria are found in many Acacia species and other woody species (El-Juhany et al., 2001). In Ethiopia, charcoal production heavily depends on acacia species for the quality they constitute of (Tinsae et al., 2012).

3.3. The Charcoal Makers

Scherr et al. (2004) estimated the world's forest-dependent poor to range from 1 billion to 1.5 billion. Vira and Kontoleon's (2010) review of people's dependency on forest shows that the poor tends to depend excessively on relatively low value or 'inferior' goods and services of forest products. A typical poor rural household harvest wood and make charcoal for sale, gather fruits and roots for food, medicinal plants, and building materials for domestic use. Almost all forest resource-dependent households are in the category of 'poor' people without land or any other capital to support their livelihoods. In Nepal, according to Pokharel (2011), poor households who are dependent on common forest resources are either landless or have small piece of land. Similarly, in Zambia, out of a total of 45,500 people engaged in the fuel-wood industry full-time, about 41,000 are engaged in charcoal production and are all from among the rural poor community (Syampungani, et al., 2009).

The Ethiopian situation is not different. Most of the charcoal entering Ethiopian cities and towns are from dry woodlands where rain fall is scarce with frequent drought and food insecurity. Beside the high vulnerability to natural hazards, communities commonly suffer from poor infrastructure and shortage of public services. Forest resources are particularly important for these poor communities who reside in the dry woodlands of Africa, what Syampungani et al. (2009), called "poverty hotspots". The field assessment also confirmed that charcoal makers in the major charcoal producing centers of the country are those without much alternative means of livelihoods.

The survey also shows that there is a clear division of gender in the charcoal business. During field visits to the major charcoal producing centers (Gewanie and Bilatie), no women were found engaged in charcoal making activities. Moreover, all distributors and about 67% of the wholesalers are found to be men. Women are often involved in retail business in Addis Ababa and accounted for about 64%. Charcoal distribution and marketing appears to be dominated by the young men (about 47% fall within the 20-30 age class). The level of education of most respondents is found to be low as most were either illiterate or attended only primary school (Table 11).

3.4. Charcoal Making Technologies

(a) Traditional Kiln

Charcoal is produced using both traditional and modern techniques. The efficiency and the yield obtained are determined not only by the type of technology used, but also by the qualities of the biomass fed, as well as the producer's skills. Earth mound and earth pit kilns are among the most widely

19

used traditional technologies for charcoal making in developing countries, including Ethiopia (FAO, 1987; Yisehak and Duraisamy, 2008).

Various studies conducted on charcoal making processes in traditional kiln in developing countries (Foley, 1986; Girard, 2002; Kammen and Lew, 2005), and the description provided by charcoal makers in Langano (Oromiya), Gewanie (Afar) and Bilatie (SNNPR) revealed similar charcoal making processes. Wood is first gathered and cut to manageable sizes, and placed in an underground (earth pit) or above ground (earth mound) kiln. Then, the kiln is fired/ignited and the wood heats up and begins to carbonize. The kiln is mostly sealed, although a few vents are initially left open for steam and smoke to escape. The production process may take a few days, a week or more depending on wood type, labor input, kiln size and producer's skill. When the process has ended and the kilns are cooled down, they are opened or dug up and the charcoal is harvested. The resulting charcoal resembles smaller, lighter pieces of blackened wood. About half of the energy in the wood is typically lost in the process but the charcoal produced has higher energy content per unit mass than firewood.

Uhart (1975) observed charcoal making in Ethiopia using earth kilns in the 1970s. The volume of the kiln, as he noted varied with the scale of commercial operation in a given area. The largest earth kiln he observed had a physical size of 10 x 10 x 5m in height with a capacity of holding 100 m3 of fresh wood. In Modjo district, about 80 km east of Addis Ababa where relatively large scale commercial charcoaling has been in progress since 1910, Uhart observed a charcoal producer who employed about 50 workers and operated 10 to 15 earth kilns at the same time. The method of production consists of earth-mound type.

In Ethiopia, charcoal is commonly produced using the traditional earth kiln method–earth mound kiln and earth pit kiln; earth mound kiln being the most frequent method with an efficiency of 10-15% (at wet weight basis) (Yisehak and Duraisamy, 2008). Earth mound kiln is the least efficient method; and charcoal produced by this type of kiln could also be easily contaminated with soil and other foreign particles.

Figure 5: Commonly used charcoal production steps using earth mound kiln in Ethiopia

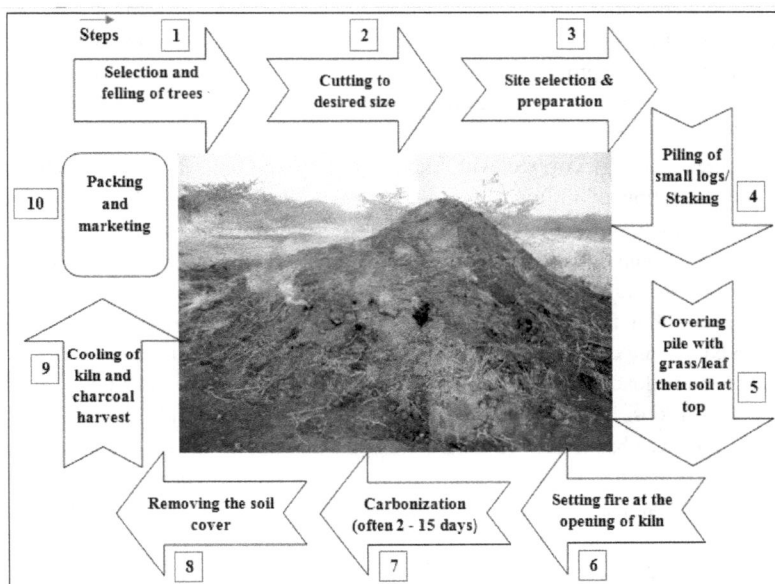

SOURCE: Compiled from Field Survey, 2012

The total carbonization period depends on the size of the mound ranges from a few days to weeks. The set of activities an Ethiopian charcoal maker follows to produce charcoal using earth mound kiln are shortly shown in Figure 5.

Gudeto's charcoal making description:

Gudeto, who is about 63 years old, is a resident of Keraro Kebele, Arsi-Negelle district, Oromiya Regional State. He began charcoal making during the Imperial period and continued all through the *Derg* time. He was interviewed in *Keraro* where he was recorded giving the following description:

We, Oromos did not make charcoal traditionally. We did not know how to make charcoal. Tree felling was a taboo. If we have to cut trees it was for the construction of our houses or beehives. The skill of charcoal making was introduced in our locality by the coming of outsiders as land owners in 1950s. The owners of the land used to bring charcoal makers from nearby towns. We

---

learnt the techniques and started to make charcoal as well. In charcoal making, the first thing is to look for suitable tree(s), often medium size trees. The fallen tree is chopped into 0.9 to 1 m size. Until the wood dries to a certain limit, we start preparing the pit. The size of the pit depends on the human labor available. It is often in a size to produce twenty to thirty sacks of charcoal. On the floor of the prepared pit we lay dry branches or leaves; at the middle of the pit a four to five feet log is positioned vertically around which the smaller logs are put in circles. Then the mound is covered with leaves, grasses and soil. The latter helps to restrict the inflow of air (oxygen) in to the wood. Two or three small openings are left at the bottom of the mound to start the fire. The most important process is to check the amount of air reaching the wood. If air is left unchecked the wood will burn down and turn into ashes. Depending on the amount of wood, this process would take four to five days to turn into charcoal. Once completed, the soil is removed and the charcoal is left to cool after which we put it into sacks and make it ready to be loaded on a car. A sack of charcoal was sold for birr 3.00 in 1950s and early 1960s. Half of the sale goes to the land owner. It did not bring much money; it was a backbreaking work done in the absence of any other alternatives. Charcoal burning destroyed our environment, prevented rain and the land has dried. Charcoal making is now made illegal; anyways, there are no trees left.

Through field observation, it was learnt that traditional charcoal making requires less capital and can easily be constructed using locally available materials. However, carbonization in such kilns takes longer time and the process requires close attention, and there can be contamination with ash, sand and mud in the course of the process. Ventilation, especially with earth pit kiln, may also be difficult to control and often carbonization is incomplete, producing only low quality charcoal. According to Kwaschik (2008), the efficiency of most traditional kilns used in African countries varies only between 10 and 25%; while Yisehak and Duraisamy (2008) put the estimate (for Ethiopian kilns) in ranges from 10-15% (at wet weight basis). Table 5 summarizes the features and efficiency of some traditional and improved kilning technologies.

Table 5: Comparison among different charcoal making techniques

Charcoal making technique	Capital investment	Labor demand	Suitability for controlling and follow up	Productivity/efficiency	Product quality
Earth pit	Low	Very high	Very difficult	15-25	Very different
Earth mound (staked horizontally or rectangular)	Low	High	Difficult	15-25	Different/ better
Earth mound (staked circular or rectangular)	Low	High	Difficult	15-30	Different/better than pit charcoal
Casamance	Low	High	Relatively simple/easier	25-30	Good
Charcoaling using barrel materials	Low	Low	Simpler/Easier	25-30	Good
Portable metal kiln	High	High	Very Simple	25-35	Good

SOURCE: FAO, 1993

(b) Improved Kilns

In improved kilns, carbonization is faster and more uniform and enables to have higher quality of charcoal and efficiency up to 30 % (Kwaschik, 2008; Nketiah, 2008). Nevertheless, improved kilning technologies do involve substantial investments; thus have limited applicability in Africa. The Casamance kiln is an example of these kilns because despite its efficiency, its adoption rates in Mozambique was still low (Kwaschik, 2008). Available charcoal making technologies tested in Ethiopia are: earth mound, Casamance, metal kiln and drum charring units (Yisehak and Duraisamy, 2008; Abebe, 2004).

Casamance kiln (improved earth mound):

The Casamance kiln is an improved version of earth mound technique where barrel (drum) is welded to serve as smoke outlet. It is covered with earth and leaf material with holes at the base and a chimney made from 3 oil drums (200 liters) is introduced (see Figure 6). The fourth drum is welded into this piece at a right angle. The fifth barrel is used to collect tar at the bottom of the chimney.

Opposite the direction of the wind a chimney is inserted into the triangular duct of the kiln (Yisehak and Duraisamy, 2008).

Figure 6: Typical Casamance kiln in Senegal (left) and Ethiopia (right)

SOURCE: GIZ, 2012; Yisehak and Duraisamy, 2008

(ii) Portable Metal Kiln:

These types of metal kilns (Figure 7) are made from metal sheets or manufactured from used 200 liter oil drums. It has a diameter of 1.70 meter and height of 0.90 meter. It has a capacity of about 2 m3 of raw material per charge. The efficiency of these kilns was found to vary from 28 -33% depending on operator skill and experience (Yisehak and Duraisamy, 2008).

Figure 7: Carbonization of Prosopis using metal kiln

SOURCE: Yisehak and Duraisamy, 2008

24

(iii) Kiln made of old barrel materials:

This type of kiln can be made from one or two old barrel (drum) materials. This method is cheaper and helps to produce a better quality charcoal (compared to other traditional methods), but it enables produce only a small amount at a time (Abebe, 2004).

(iv) Subri–Fosse:

This is an improved version of traditional charcoal making which combines pit based traditional charcoal making with old iron metals to cover it (see Figure 8).

Figure 8: A semi-permanent low-cost "Subri-fosse" type metal kiln in Madagascar designed to carbonize off-cuts from sawmilling

SOURCE: Girard, 2002, (Kenya)

According to the Rural Energy Resources Development Bureau of Amhara Regional State (2011), the aforementioned improved kilns are currently being introduced by the agency into the region and attempts have already been started to test them. But, they have got their own pros and cons; summary of the pros and cons is given in the table below (Table 6).

Table 6: Advantages and disadvantages of improved kilns

Type of kilns	Advantages	Disadvantages
Casamance (improved earth mound kiln)	- Increases productivity by 10 -15%; reduced time needed to make charcoal by 50% when compared to traditional earth mound kiln; - Compared to other modern technologies, it is less costly; and it is manageable	- Like traditional method, it uses soil to cover the wood. - Hence, takes time to be cooled, labor intensive to separate charcoal from the soil and other debris since they are mixed
Portable metal kiln	- Reduces labor, improves productivity by 25-35% compared to traditional method, - Manageable, takes short time and gives quality charcoal	- It's too costly compared to other technologies, requires investment
Kiln made of old barrel materials	- Gives good productivity compared to traditional method; can give about 25-30% charcoal of the wood used; - Manageable and takes short time; gives quality charcoal and is cheaper	- Difficult to make charcoal from large amount of wood at a time
Subri-fosse	- Manageable and simple to follow up; - Improves productivity of traditional techniques up to 10% and is cheaper	- Since it is produced in pit, its quantity and quality is governed by soil moisture content; - Requires pit making at every production time in different places

SOURCE: Abebe, 2004

3.5. Production of Charcoal Briquettes

A briquette is a block of flammable matter used as fuel to start and maintain a fire. Charcoal briquettes are products intended to substitute wood charcoal. They are either made from charcoal residues left over at charcoal lump production sites or they are made from biomass that are not suitable for the manufacture of wood charcoal. Residues from agriculture and forestry, for example, bagasse, coffee husks and saw-dust, are valuable sources of raw material used to produce briquettes (GIZ, 2012).

Briquetting is one of the several compaction technologies to form a product of higher bulk density, lower moisture content, and uniform size shape (Wondwossen, 2009). The process entails many steps. First, the material is dried before it is converted to charcoal in a charring kiln. The carbonized biomass is then mixed with water and locally-available binders such as starch, gum arabic, molasses, clay and others. Finally, the mixture (powdered charcoal and binder mixture) is pressed into briquettes. Piston and screw presses are the most widely used technologies where as in the developed countries roll presses are more common. After a subsequent drying step, the briquettes will develop the required strength and stability. Forest and agricultural waste charcoal briquettes, with about 20% of clay, produce about 12 MJ/kg (GIZ, 2012).

In Ethiopia, production of charcoal briquette from different agricultural and/or forest wastes has already been started by the Ethiopian Rural Energy Development and Promotion Center (EREDPC). Biomass wastes from bamboo, prosopis, cotton stalk, Chat (Khat) stem and coffee husks have shown promising results (Table 4). The overall production process is shown in Figure 9.

Figure 9: Charcoal production and briquette making technology process chain

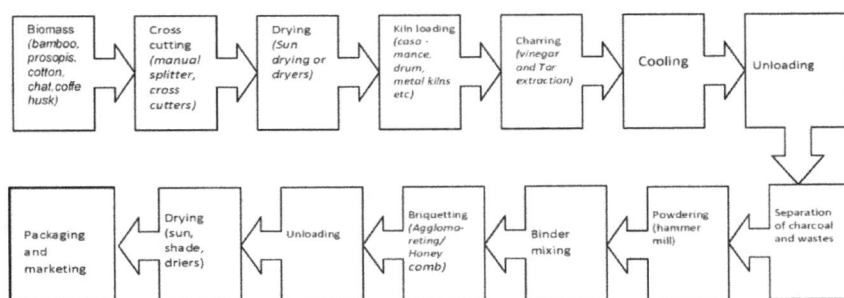

SOURCE: Yisehak and Duraisamy, 2008

(Note: The upper part of the chain is the process which is being developed by the Ethiopian Rural Energy Development and Promotion Center (EREDPC) for converting any hard biomass into charcoal and to make briquettes)

Various types of charcoal briquettes can be produced; beehive briquettes[8] (Figure 10a) and agglobriquettes[9] (Figure 10b) being the most common. Beehive briquettes require a specially made stove and are suitable for activities that require continuous heat for relatively longer hours than is required from common charcoal stoves. Agglobriquettes are produced by agglomeration-a method of size enlargement by gluing charcoal powder particles using binder in a rotating pan agglomerator (Figure 10b) (Yisehak and Duraisamy, 2008).

Figure 10: Charcoal briquetting; (a) beehive briquettes, (b) agglobriquettes produced from Bamboo waste

(a)   (b)

SOURCE: Yisehak and Duraisamy, 2008

Although, agricultural residues are employed for diverse uses in Ethiopia, there are always wastes from agriculture and forest industries that can potentially be used for briquette making. The other advantage of briquette is that it has little smoke, with less impact on human health and the environment. Briquette charcoal is, thus, viewed as an advanced fuel because of its clean burning nature and the fact that it can be stored for long periods of time without degradation (Wondwossen, 2009). According to the same author, unlike wood charcoal, briquette charcoal making is simple and safe (Table 7).

---

[8]Beehive briquettes (produced through beehive briquetting) are cylindrical in shape (13 cm in diameter and 8 cm long) and have a number of holes (usually 19 longitudinal holes of 13 mm diameter). A single beehive char-briquette will weigh about 400 to 500 grams (source: Yisehak and Duraisamy, 2008).

[9]Agglobriquettes (produced by agglomeration) are spherical and have a diameter which varies between 20 and 30 mm. The charcoal powder is shaped in to briquettes by using a pan agglomerator; the production capacity of an agglomerator is between 30 and 50 kg/hr (source: Yisehak and Duraisamy, 2008).

Table 7: Comparison of wood charcoal and briquette charcoal making processes

Briquette Charcoal	Wood Charcoal
- No need of digging a ground to prepare shallow pit of charring (low production cost)	- Digging (higher production cost)
- Mobile (Its mobility allows working at a spot of harvesting, farmstead and anywhere)	- Not mobile
- It is not fire hazardous	- Sometimes it is fire hazardous
- It is safe in the view of health factor	- It is not safe

SOURCE: Wondwossen, 2009.

Besides, briquette charcoal is of better quality compared to wood charcoal; it is smokeless fuel as the smoke is assumed to disappear during carbonization, and burns longer (see Table 8).

Table 8: Comparison of briquette charcoal and wood charcoal

Briquette Charcoal	Wood Charcoal
- Smokeless	- Smoke
- It exhibits faster heat release and greater heat value	- Less heat release and smaller heat value
- Reduce impact of deforestation	- Enhance deforestation impact
- It burns longer (2-3 hr)	- It burns for short time (1-2hr)

SOURCE: Wondwossen, 2009

The briquetting technology has, thus, a great potential for converting waste biomass into a superior fuel for household use, in an affordable, efficient and environmentally friendly manner. Consequently, it can be concluded that charcoal briquetting using agricultural wastes and other materials can be a viable alternative to wood charcoal owing to its multi-faceted benefits from economic, health and environmental point of view.

Figure 11: Distribution of the dominant land cover types by region (ha)

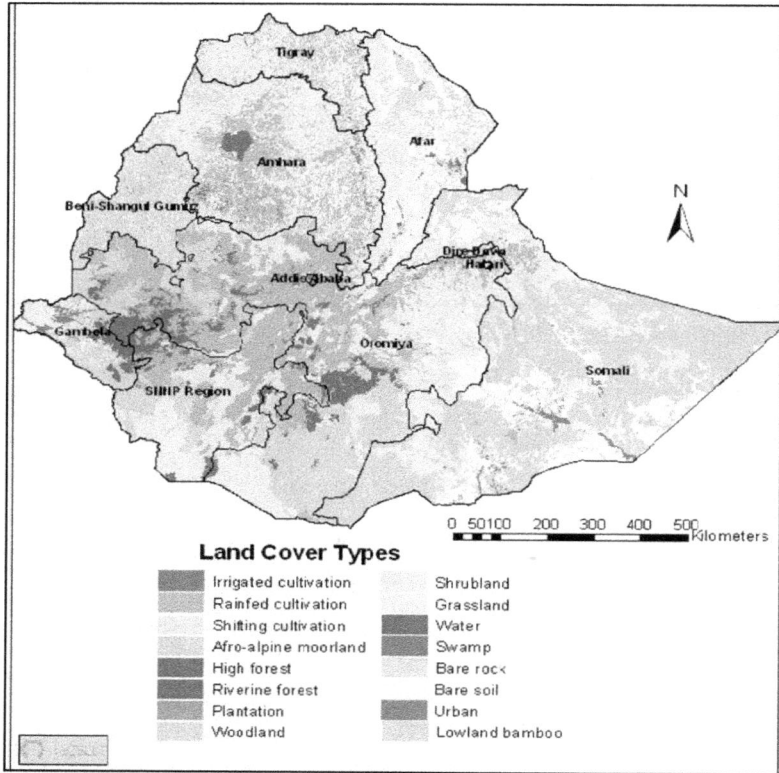

SOURCE: MoARD, 2005

3.6. Charcoal Producing Areas and the Preferred Tree Species

Malimbwi et al. (2010) noted the tropical dry-forests and woodlands to be the main source of fuel-wood consumed in both rural and urban areas throughout SSA. Kwaschik (2008) also revealed that the Miombo woodlands, which belong to the dry tropical woodlands, are the main vegetation resources to produce charcoal in various East African countries: such as Mozambique, Malawi, Tanzania and Zambia.

Similarly, in Ethiopia, the acacia-dominated dry-woodland and shrubland areas (Table 9 and Figure 11) which cover about 60% of the total landmass of the

30

country (WBISPP, 2004), constitute the largest source of wood for the bulk of charcoal coming to urban centers.

They have been over-exploited freely for decades as the property rights on these resources are loosely established and there is little control over the resource base.

Table 9: Woodland areas of Ethiopia

Region	Area (ha)	% of total woodland
Oromiya	9,823,163	34%
SNNR	1,387,759	5%
Gambella	861,126	3%
Amhara	1,040,064	4%
Tigray	254,455	1%
Benishangul-Gumuz	2,473, 064	8%
Afar	163,667	1%
Somali	13,199,662	45%
Total	29,202,960	

SOURCES: WBIPP, 2000; unpublished

According to Zerihun and Mesfin (1990), the Rift Valley vegetation is an important source of charcoal produced for the nearby towns and Addis Ababa. Respondents have also confirmed that Afar region and areas in the vicinity, including Awash and Langano, are among the important sources of charcoal to Addis Ababa. The amount of charcoal that comes from plantation forests is not known. Although survey for Addis Ababa indicates that the charcoal entering the capital through Sebeta gate is known to be produced from Eucalyptus, the bulk of charcoal comes from either acacia species and/or the invasive species– Prosopis juliflora. According to respondents in the visited areas, and also in accordance with several other studies[10], the various acacia species are most

---

[10]See Zerihun and Mesfin (1990); El-Juhany et al. (2001); Damascene (2005); Mugo and Ong (2006); Falcão (2008); Kwaschik (2008); Yisehak and Duraisamy (2008); PA (2010); and Tinsae et al. (2012)

popular trees for charcoal in Ethiopia, that includes Acacia tortilis, A. mellifera, A. senegal and A. seyal. There are also many other tree species[11] reported to be used for charcoal making.

3.7. Charcoal Production Trends

Charcoal, which covers about 80% of urban households' energy needs in Africa, remains one of the prime sources of energy in the continent, particularly in SSA. And, yet it will remain the main cooking fuel for most people in the region's towns and cities for the foreseeable future because it is accessible and affordable (Mugo and Ong, 2006). With population increase, urbanization, and economic growth, the demand for energy is expected to grow. As the modern energy sources are still beyond the reach of the majority of people in developing countries, dependence on biomass fuel is expected to continue (AREAP, 2011).

According to Adam's (2008) assessment, Kenya consumes about 2.4 million tons of charcoal annually, while Zambia uses about 1 million tons. The World Bank's (2009) assessment from 2001-2007 also showed that the number of households in Dar es salaam, Tanzania cooking with charcoal grew from 47% to 71%, while the use of LPG declined from 43% to 12%.

The global production of wood charcoal was estimated at 47 million metric tons in 2009, and increased by 9% since 2004. This increase is strongly influenced by Africa, which produces about 63% of the global charcoal production. Charcoal production boosted in the continent by almost 30% since 2004, thus, extended Africa's global lead (FAO-STAT, 2011). Consequently, the escalating rate of wood charcoal production, particularly in developing countries, will continue to pose severe threats on the remnant woodland resources.

---

[11]For example, Azadirachta indica; Balanities aegyptica; Boswellia sp; Combretum sp; Commiphora sp; Dichrostachys cinerea; Diospyros mespiliformis; Eucalyptus sp; Ficus sp; Grewia sp; Grevillea robusta; Hagenia abyssinica; Jacaranda mimosifolia; Lantana camara; Olea europea; Prosopis juliflora; Senna *senguiana; Sesbania sesban; Syzygium guineense; Terminalia brownii; Tamarindus indica; Ximenia americana; and Ziziphus sp.*

Figure 12: Top ten wood charcoal producing countries in the world

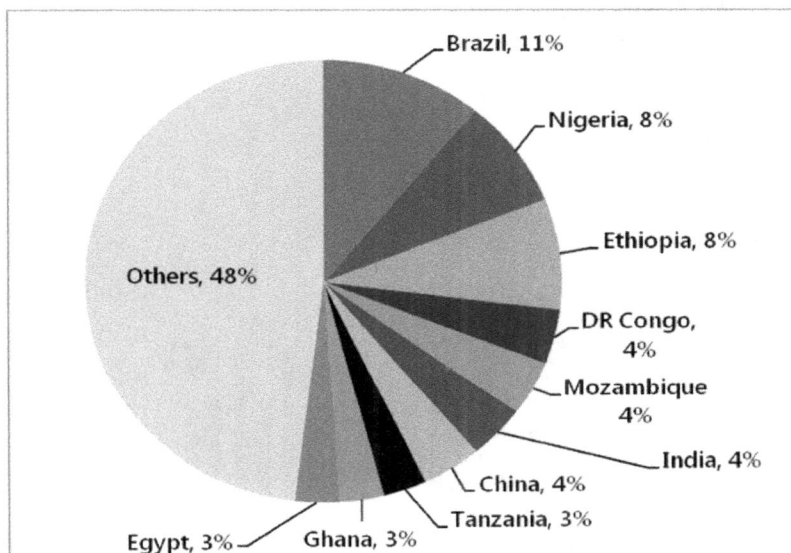

SOURCE: FAOSTAT, 2011

Among the top ten wood charcoal producing countries in the world, Brazil, with the largest forest resource in the world, stood first; while Nigeria and Ethiopia are second and third (Figure 12). The remaining seven countries are: Democratic Republic Congo, Mozambique, India, China, Tanzania, Ghana and Egypt. Globally, at least 19 million tons of wood are consumed each year to produce charcoal by thousands of traditional charcoal producers, and is even set to increase at faster rates in the near future (FAO STAT, 2011).

Figure 13: Production of wood charcoal in three East African countries
('000 tons)

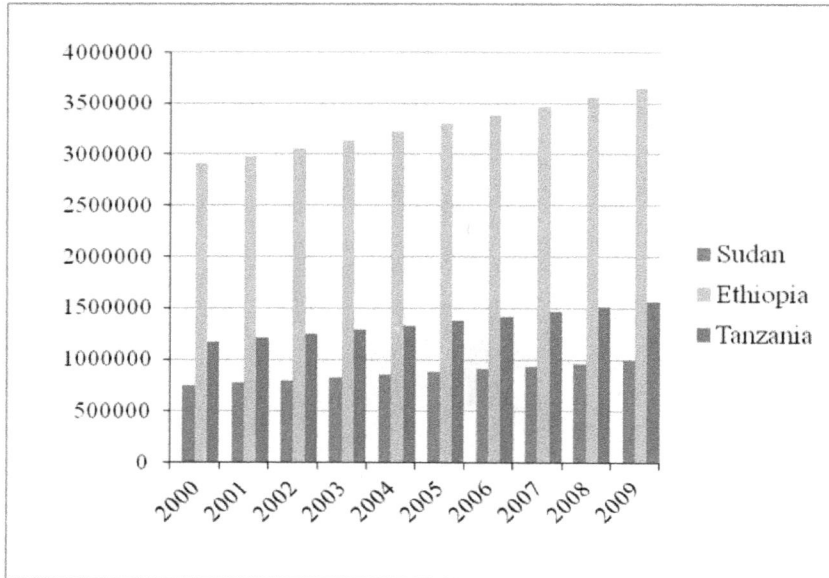

SOURCE: FAOSTAT, 2011

The trend of charcoal production among east African countries (e.g. Ethiopia, Sudan and Tanzania) showed rapid increment since 2000. Between 1995 and 2005, Ethiopia produced an estimated amount of 3,320,535 tons of charcoal. In 2009, the volume of charcoal produced in Ethiopia increased to about 3.6 million tons (Figure 13).

4. Charcoal Supply, Marketing and Consumption in Selected Cities and Towns

4.1. Introduction

Household energy use can, generally, be categorized as traditional (including agricultural residues and firewood), intermediate (charcoal and kerosene) or modern (LPG, biogas and electricity) (Msuya, 2011). In developing countries, particularly SSA, energy consumption is still low and limited almost exclusively to biomass fuels: fire-wood, charcoal and other organic wastes (Girard, 2002; Malimbwi et al., 2010).

Table 10: People relying on wood-based biomass (millions)

	2004	2015		2030	
	A	A	B	A	B
Sub-Saharan Africa	575	627	741	720	918
North Africa	4	5	4	5	4
India	740	777	863	782	780
China	480	453	393	394	280
Rest of developing Asia	645	692	688	741	709
Latin America	83	86	85	85	79
Total	2527	2640	2774	2727	2770

SOURCE: (A) World Energy Outlook 2006 (OECD/IEA, 2006); (B) World Energy Outlook 2010 (OECD/IEA, 2010)

Unless alternative household energy sources are devised, with increasing population growth, the number of people dependent on biomass will increase to over 2.6 billion by 2015 and to 2.7 billion by 2030 (Table 10). While the use of biomass fuels in China, India, and much of the developing world has peaked or will do so in the near future, SSA's consumption will either remain at very high levels or even grow over the next few decades (AREAP, 2011).

With mounting urbanization, African populations (e.g. See the case of Mali in Figure 14) are increasingly shifting from firewood to charcoal for domestic cooking and heating. The increasing shift from firewood to charcoal for

35

domestic use could, among others, be attributed to the fact that charcoal is easier to transport, efficient and produces a steady heat with little or no smoke compared to firewood (Girard, 2002).

Figure 14: Use of wood versus charcoal as fuel in Bamako, Mali

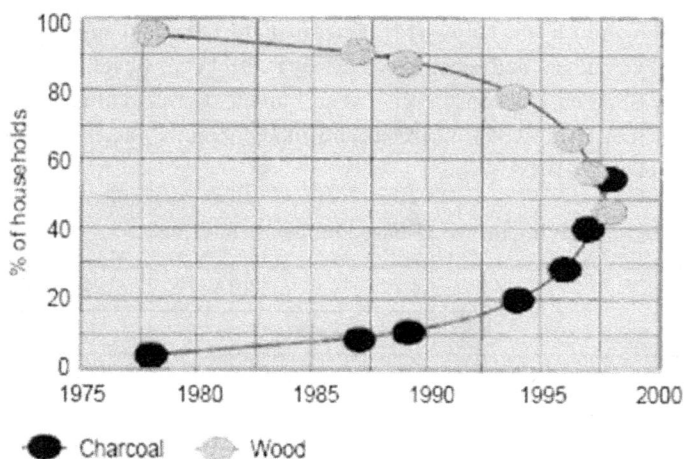

SOURCE: World Bank, 2000

Charcoal is a convenient and accessible energy source for cooking at all times and at a reasonable cost. In addition, charcoal trade offers income generation opportunities for many people in the urban areas, through small scale retail businesses mostly run by women who sell charcoal in the urban roads. All these factors along with the absence of affordable and convenient modern alternative energies rendered charcoal to be consumed at higher rates among urban areas. The switch to charcoal is reported to continue at a rate between 4% and 10% per year. At the same time, the further switch to more convenient fuels on the energy ladder is expected to be hampered by high oil prices (Girard, 2002; PREDAS, 2008).

Amount and type of energy consumption is closely related to the level of a given country's economic development (UN-Energy, 2005). A World Bank survey in 45 cities in 12 countries between 1984 and 1993 showed that a decrease in the use of fuel-wood and a shift to modern sources are related to accessibility, and improvement in incomes and favourable government policies (Waddams, 2000).

36

As a result of household income differentiations, there is a sort of ladder of energy sources in the urban areas in SSA: from fuel-wood at the bottom, through charcoal, kerosene and gas, to electricity at the top (See Figure 15).

Figure 15: The energy ladder of cooking fuels

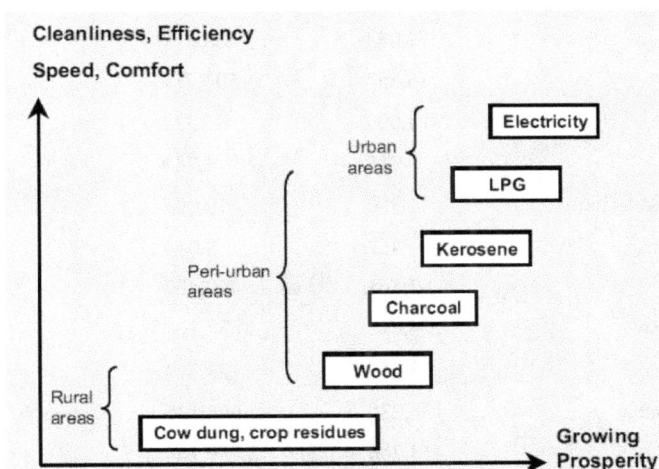

SOURCE: PREDAS, 2008

Households, generally, ascend this ladder as their income increases. Along the ladder, an increment in cleanliness, efficiency, cost and convenience is shown in the direction of increasing prosperity (Figure 15). In some cases where there is fuel-wood scarcity (e.g. in the case of Ethiopia), one finds cow-dung at the lower bottom of sources of fuel. This shows that majority of the rural poor in developing countries use inefficient and relatively more polluting energy sources, and most often are forced to engage in ecologically damaging activities.

Charcoal is not used as a fuel in most areas of rural Ethiopia. Given its convenience and accessibility, charcoal is essentially an urban fuel across the whole country. The estimated total amount of charcoal consumed by each region is shown in Table 11. The estimates show that Addis Ababa is by far the largest consumer of charcoal owing to the very high urban population. Oromiya, Amhara and Somali regions are the next highest consumers. This is indeed a reflection of the relatively higher rates of urbanization.

Table 11: Total annual urban consumption of charcoal and its wood equivalent* by Region (tons)

Region	Charcoal	Charcoal as wood	As % of Ethiopia
Tigray	38,307	255,380	9%
Amhara	72,033	480,217	17%
Oromiya	75,557	503,714	18%
SNNP	13,099	87,324	3%
Afar	715	4,764	0%
Benishangul-Gumuz	1,380	10,306	0%
Gambella	452	3,014	0%
Somali	67,561	450,407	16%
Dire Dawa	4,825	32,168	1%
Harari	3,557	23,714	1%
Addis Ababa	136,220	908,135	33%
ETHIOPIA	413,706	2,759,144	

* Assuming 15% efficiency in charcoal transformation (wood to charcoal conversion)

SOURCE: MoARD, 2005

Charcoal consumption and marketing study for Addis Ababa, and charcoal issues on the selected major towns and cities, and charcoal producing areas are presented in the following pages.

4.2. The City of Addis Ababa

The city of Addis Ababa has been expanding physically mainly due to natural population increase and internal migration (Minwuyelet, 2005). Currently, the city occupies 54,000 ha of land. Although the official statistics put the city's population at 2.8 million, growing at the rate of 3.7% per annum (CSA, 2008), most authorities on the subject tend to hold the view that the city's population is underestimated. And estimate the city's population as not less than 3.5 million (Aynalem Adugna, 2008, unpubl.).

Table 12: Characteristics of respondents involved in charcoal business in Addis Ababa, Ethiopia

Variable	Number (%) of Respondents						Total (%)
	Wholesalers (N=100)%		Retailers (N=95)%		Distributors (N=7)%		
Age							
20-30	58	58.0	34	35.8	3	42.9	47.0
30-40	32	32.0	41	43.2	3	42.9	37.6
>40	10	10.0	20	21.1	1	14.3	15.3
Sex							
Female	33	33.0	61	64.2	0	0.0	46.5
Male	67	67.0	39	41.1	7	100	55.9
Education							
No school	26	26.0	56	58.9	0	0.0	40.6
Primary	43	43.0	28	29.5	5	71.4	37.6
Secondary	30	30.0	10	10.5	2	28.6	20.8
Higher	1	1.0	1	1.1	0	0.0	0.99

N= number of respondents

SOURCE: based on data from field survey, 2012

4.2.1. General Features of Respondents

In order to assess the supply, demand, marketing and consumption and other related issues of charcoal, the study, correspondingly, looked into four groups of respondents: distributors, wholesalers, retailers and consumers.

4.2.2. Views of Respondents on Charcoal Supply, Marketing and Consumption

Nowadays, there is a much heightened demand of charcoal among the urban dwellers due to the fact that charcoal is affordable and accessible household energy source as compared to other modern alternative energy sources. Besides, it is easier to start the charcoal business as well as the income from the business is promising. As a result, the charcoal market is expanding and attracting more people. For example, about 71.4% of the distributors, 74% of the wholesalers and 80% of the retailers surveyed reported that they have joined the charcoal business only in the past five years (Table 13). At the same time there could be people who could leave this business.

Table 13: Respondents' level of dependence and years of engagement in charcoal business

Variable	% of Respondents			Total
	Distributors (N=7)	Wholesalers (N=100)	Retailers (N=95)	(N=202)
Number of Years:				
<1	28.6	14.0	16.8	15.8
1 – 5	71.4	74.0	80.0	76.7
6 – 10	0.0	10.0	3.2	6.4
>10	0.0	2.0	0.0	0.99
Entire dependence:				
Yes	100	42.0	20.0	33.7
No	0.0	58.0	80.0	66.3

SOURCE: based on data from field survey, 2012

Charcoal business is not the sole means of livelihood for most of the respondents (66.3%) surveyed; they are engaged in the business looking for additional income (Table 13). Yet, the business is a principal source of income for the

40

distributors. As illustrated in Figure 16, most respondents indicated that they entered into charcoal business either due to lack of job opportunity or for its profitability. A minority of respondents took-over the business from their parents.

Figure 16: Reasons to be in charcoal business (% of respondents)

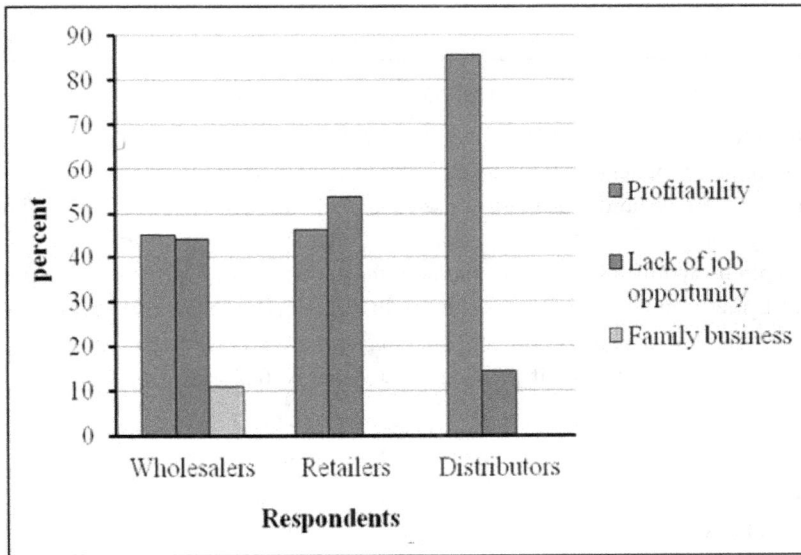

SOURCE: based on data from field survey, 2012

According to the survey, charcoal is reported as the main household energy source for most households (about 73%) in Addis Ababa (Table 14).

Table 14: Type of energy sources used by households (% of respondents)

Energy Source	%
Charcoal	72.9
Electricity	18.9
Fire-wood	2.7
Gas	5.4

SOURCE: based on data from field survey, 2012

Although the degree and type of consumption may vary, most urban household, be it poor or better-off tend to use charcoal as source of energy for varied purposes. Households primarily use charcoal for cooking, coffee making, heating, roasting of maize, and ironing of clothes, among others. There is a perception among household consumers that cooking with charcoal makes the food tastier compared to that cooked with fire-wood. It was also reported that consumers preferred dense (heavy) charcoal and are willing to pay more for this type of charcoal because it provides more energy (calories) and burns longer with little smoke.

Households also stated that their charcoal consumption has gone up in the last few years in connection to price inflation of other alternative energy sources, especially kerosene and LPG. Charcoal use during weekends and festivities is high as they had more cooking to do and engage in social get-together. Respondents (about 78% of the surveyed consumers) observe that the quality of charcoal entering Addis Ababa is declining with time. The supply (about 51%) of charcoal was also reported to be decreasing. This could be attributed to the on-going degradation of the acacia forest resource base. But then, the price of charcoal has shown an increase owing to its much heightened demand among the urban households. All of the surveyed consumers confirmed the escalated increment in price of charcoal, while majority (about 76%) has also indicated increment in demand of charcoal in the past few years.

With the exception of some who perceived similar demand of charcoal, most respondents confirmed that there is evident seasonal variation in supply, demand and price of charcoal, which is reported to escalate during kiremet (the long rainy season from July to September) season.

Table 15: Main problems facing charcoal business as prioritized by respondents

Problems/threats	Priority rank given (% of respondents)		
	1st	2nd	3rd
Price hike	39.6	31.7	11.9
Shortage of supply	38.1	33.2	18.3
Market problem	7.9	18.8	50.9
Problems at checkpoints	7.4	4.9	8.9
High taxation	1.9	2.9	0.99

SOURCE: based on data from field survey, 2012

The charcoal business is currently confronting enormous challenges along its various stages starting from production through marketing to consumption. As prioritized by the respondents, price hike and shortage of supply are among the main challenges facing the charcoal business (Table 15).

4.2.3. Charcoal Inflow

Charcoal inflow survey was conducted in August 2012 to know the amount of charcoal entering into Addis Ababa through its major inlets. For this purpose, a 24–hours counting and recording of charcoal trucks and other vehicles with sacks of charcoal was conducted for four days in each of the five major gates of the capital.

Figure 17: Charcoal inflow to Addis Ababa city through five main gates

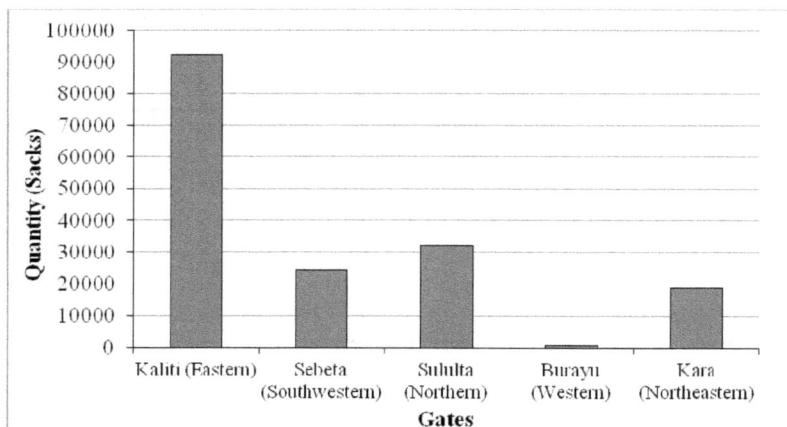

SOURCE: based on data collected between August and September, 2012.

Most of the charcoal (about 55%) entering Addis Ababa comes through the eastern gate (Kaliti), i.e., Awash, Gewanie and Bure-Dimtu in Afar, by ISUZU medium duty trucks. Charcoal coming through Sebeta road is from localities along the road to Jima. Charcoal from the Abay gorge comes through the northern gate of Sululta. The western Burayu entrance receives charcoal from Ginchi, Enchini and Wolkiete, while charcoal coming to Addis through the northeastern entry point of Kara is sourced from Shewa-Robit, and Dessie area. Accordingly, about 42,045 sacks[12], each weighing about 35 kg on average, were supplied to the city in a day (Figure 17), suggesting an equivalent of 537,124.875 tons of charcoal per annum. According to the survey, even though a number of vehicles including heavy load trucks, pick-ups and automobiles carry charcoal into the city, most of the charcoal entering into Addis Ababa is conveyed by means of ISUZU medium duty trucks.

---

[12]This charcoal supply seems bigger as compared to other survey results probably because the survey: (i) was carried out during the cold kiremet season where charcoal demand is at its peak; (ii) counting considered wide range of transportation means (Isuzu medium duty trucks, Heavy load trucks and Automobiles); and (iii) was conducted at the time where charcoal was freely supplied to the city from the Afar region, as the region allowed charcoal burning as a way of controlling the spread of the invasive plant called Prosopis.

4.2.4. The Charcoal Trade Chain

The charcoal transported to the city through the five gates is delivered to depot owners (distributors) stationed at different corners of the city. Consumers and retailers could buy charcoal directly from a distributor. The summary of the possible channels of charcoal supply to urban consumers in Addis Ababa is illustrated in the diagram below (Figure 18).

Figure 18: Charcoal supply chain to Addis Ababa city, Ethiopia

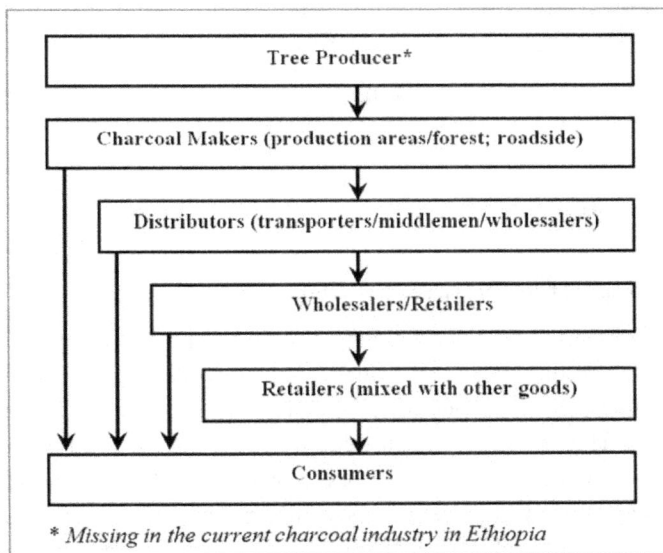

SOURCE: Compiled based on data from field survey, 2012

While there are indirect actors whose actions either help or hinder the marketing process, the main actors directly involved along the charcoal marketing chains include producers, distributors/transporters, wholesalers, retailers and consumers.

Figure 19: Main actors and possible beneficiaries along the charcoal chain

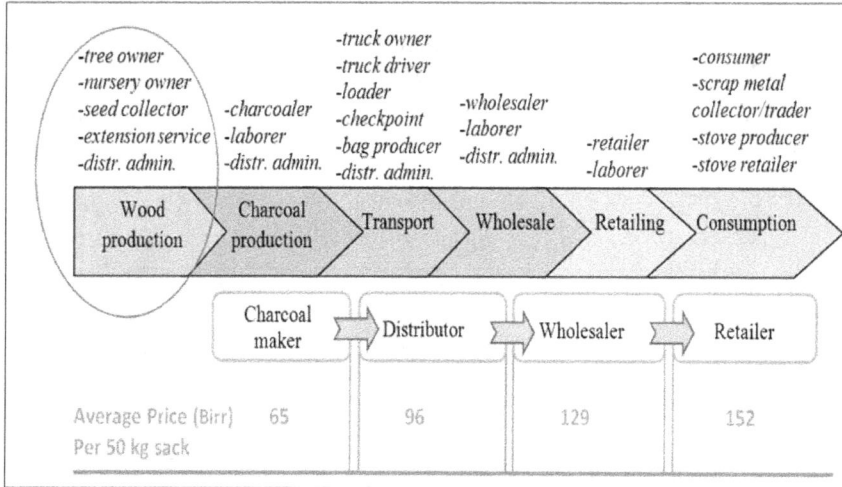

SOURCE: Compiled based on data collected from field survey, 2012.

Under healthy policy framework where charcoal is produced from a properly managed forest resources, and the trade is regulated, all actors along the market chain including those involved in tree growing would receive their share from the business (Figure 19). Nonetheless, in Ethiopia, the current charcoal production system does not take the tree resource into account. Because charcoal makers simply produce charcoal from state or communal forest resources free of charge. This practice, not only depletes the forest resource, but also distorts the market price.

As the charcoal commodity is moved from the point of production through markets to consumers, it incurs various costs: production, transportation, taxation and other informal costs (e.g. bribes and payments to brokers as reported by interviewees, loading–unloading). The business is also subject to different forms of formal and informal costs, which vary depending on distances of production sites from major market and consumption centers, mode of transport, and other related issues. Thus, it is problematic to accurately present the cost-benefit distribution of the business along its chain. Therefore, only shares of retail values (not just overall profitability) of charcoal has been extrapolated based on results obtained from interviews and/or discussions with the main actors in the charcoal business in Addis Ababa and production sites. Accordingly, prices accruing to the charcoal makers, distributors, wholesalers and retailers are estimated at birr, 65, 96, 129, 152 (Figure 19).

4.3. Mekelle (Tigray Regional State)

Tigray, the northernmost region of Ethiopia, is one of the nine regional states. According to the 2007 housing and population census, the region has an estimated population of 4.3 million, about 49% and 51% being male and female, respectively. Of the total population, only about 19% are urban dwellers. There are 49 urban centers in Tigray with a population of 2000 and above. Expectedly, Mekelle, the regional capital, is the largest with a population of 214,806, followed by Adi-grat and Axum. Another seven urban centers have a population of over 25,000, and nine more have a population size between 10,000 and 25,000. A quarter of the 49 urban centers in Tigray have a population of less than 4000 (CSA, 2008). The vegetation cover of the region is dominated by the dry-woodlands and shrublands, with little patches of forest trees mainly on sloppy and mountainous areas (WBISPP, 2004).

As is the case with many other cities and towns in Ethiopia, the urban centers in Tigray are heavily dependent on purchased biomass fuel for domestic use as the survey conducted in August 2012 showed. According to the Agricultural and Rural Development Bureau of the region, of the total energy source to the town, about 85% is from biomass, while 15% is from electricity and kerosene.

Most of the charcoal producers in the region are men farmers in the age range between 20 and 35 years. Except for the landless youth, most charcoal producers are involved in the business as off-farm activity looking for additional income. Charcoal is produced mainly from the existing woodland resources on an open-access basis; the most preferred tree species being Acacia species (e.g. Acacia etbica, locally known as "Seraw") and Olea europea (Awlie'), among others. Although most of the charcoal produced in the region is meant for domestic market, charcoal produced in western zone of the region (e.g. Humera) is illegally exported to Sudan.

Currently, the main source of charcoal to Mekelle comes mainly from, and the borders of Afar Regional State. The nearby districts (woredas) supplying charcoal to the town are Enderta, Hintalo Wajirat, Saharti Samire and Degua' Temben. Charcoal is brought to Mekelle by trucks (mainly Isuzu medium duty trucks) from the Afar region and by donkey loads from the nearby districts in the region. Within the town, charcoal is then transported and distributed by means of carts. According to the Agricultual and Rural Development Bureau of the woreda, on average, about 400 donkey loads of wood-fuel (sometimes up to 1400 donkey loads each market days) enter Mekelle through Endayesus gate; about 95% estimated to come from Enderta woreda. It is also indicated that the average selling price of charcoal in its marketing channels from producers through distributors and wholesalers to retailers is estimated at birr 60, 85, 95 and 105 per 50kg sack of charcoal, respectively.

Results of the field survey showed that the price of charcoal in Mekelle varies considerably through seasons: the production decreases during kiremt season, and drops significantly during peak farming season (e.g. autumn); as a result, the price increases. On the other hand, during the time the farmers have less workload (December to May), the supply increases; as a result the price of charcoal declines. In general terms, while there is an increasing demand and growing price of charcoal, the production showed declining trend in the last few years possibly due to engagement of the youth in other development activities related to agriculture (e.g. involvement in irrigation agriculture), and lack of supply of raw materials; as a result charcoal has been supplied to the town mainly from Afar region.

Charcoal production is, generally, prohibited in Tigray. Agricultural extension agents along with security forces, in their bid to abate the escalating rate of woodland degradation, take measures against those involved in charcoal making. In this connection the regional state government has currently designated six state forests and put strict regulations to prevent any form of fuel-wood production. However, in various parts of the region, there are illegal/informal charcoal production activities. Though most of them remained ineffective, about 182 checkpoints have been established in different parts of the region to regulate the entrance of charcoal to Mekelle town. Charcoal producers and suppliers are charged if caught before entering the town. However, the one from Afar region is legalized (licensed), thus not charged. The charcoal industry in the region entertained, mostly, prohibitive measures from various agents on collective basis: there is no a specific body responsible for charcoal production and marketing. Thus, the sector still continues threatening the remnant woodland resources. Besides its environmental and health impacts, the charcoal business is becoming a source of conflict among local communities, e.g. between locals in the borders of Afar and Tigray Regional States.

Cognizant of the presence of various illegal charcoal productions and their consequences on the forest resources, the ARDB of the region in collaboration with other concerned bodies started to build up public awareness on charcoal issues through various workshops (e.g. workshops has been conducted in western and southeastern zones of the region on charcoal making and curbing deforestation). There are also projects initiated to establish large eucalyptus plantations for fuel-wood sources, and to supply for chip-wood industry thereby supplying wood wastes to be used for charcoal making.

4.4. Awash and Gewanie (Afar Regional State)

The Afar regional state, located in northeastern part of Ethiopia is one of Ethiopia's regions which are predominantly inhabited by people who are engaged in nomadic pastoralism. The region has a total area of 100,860 km2 (Hundie, 2006). Based on the 2007 census, the region has a total population of about 1,390,273, consisting of 775,117 men and 615,156 women. About 185,135 are urban inhabitants, which accounted for about 13.32% of the total population (CSA, 2008). According to Sandford and Habtu (2006) as quoted by PFE, IIRR and DF (2010), 81% of the total population is pastoralist. The Awash National Park and a number of large scale cotton, sugar-cane, and tobacco commercial farms are located in the region resulting in the reduction of the Afar rangelands. The land is dominated by dryland vegetation of acacia tree species and grasses. The recently introduced Prosopis juliflora for the purpose of soil conservation and desertification control is invading the Afar region, creating its own landscape. The species extended its habitat faster than any other species, thus remained invasive. It has no any growth limitation–it can even grow over bed rocks and took over the pasturelands quickly and poses severe challenges to the pastoral community (Farm Africa, 2006).

In the Afar region, charcoal production has been considered as taboo; thus is not a common habit among the indigenous Afar people. Most charcoal producers in the region are still new comers from other areas–they are mostly highlanders who are involved in farming activities. However, currently, even the Afars, who once perceived it as taboo, are getting involved in the business in view of its market values and its potentials to generate additional income.

49

The production of charcoal from the invasive species–Prosopis juliflora is encouraged by the regional government as a way of controlling the expansion of the species over the pasturelands. In response to the invasive Prosopis, various GOs and NGOs started to undertake different management options to curb the expansion of the invasive species, which otherwise is hampering the livelihood of the pastoralists. Uprooting and clear-cutting, debarking, and allowing massive charcoal productions from the species are among the management options that have been tested. In 2005/06, some cooperatives were established with the aid of FARM Africa to encourage large scale charcoal production from P. juliflora. Consequently, large scale charcoal production commenced in the region, making the Afar region to be the major source of charcoal to the nearby towns and to the capital, Addis Ababa. This has been confirmed both by the field observation made in the region, and interviews of consumers in Addis Ababa.

Nevertheless, none of the attempts made to control the invasive species were effective, rather the species has good sprouting ability and even becomes more pronounced following to cutting. Even though there are still some research undertakings, none of them remained effective in controlling the invasive species–Prosopis juliflora. The encouragement of producing charcoal from P. juliflora as a way of containing its spread has, however, led to cutting of trees from other species, mainly Acacia sp. from the existing woodlands. This has increased the number of illegal charcoal producers in the name of clearing P. juliflora.

According to interviews with producers near Awash and Gewanie towns, charcoal making is chiefly the role of men with the age range between 17 and 40 years. Yet, some poor widow women involve in charcoal production to support their livelihoods.

Figure 20: Piles of Prosopis wood prepared for carbonization

SOURCE: FARM Africa, 2008.

Producers in Gewanie (Afar) stressed that charcoal making has been the most difficult trade. They found it dangerous for they face enormous risks while producing it. The production is laborious and requires constant follow up. Moreover, producers reported risks to their wellbeing associated with wild animals such as lion and hyena, and various forms of diseases including respiratory diseases. Even though producers interviewed found charcoal making risky and confirmed that they are cognizant of the environmental consequences (deforestation), they highlighted how they are forced to continue producing charcoal due to the lack of other job opportunities to support their livelihoods.

Charcoal is mainly produced from the existing woodland resources in an open access basis. Prosopis juliflora and Acacia species are among the most preferred charcoal species in the region. While quality charcoal is obtained from Acacia sp, most producers rely on P. juliflora as it requires relatively less labor of production, and at the same time there exists permission to produce charcoal from the invasive species. Given the promising income from the sector and the unrestricted access to the invasive species, the number of charcoal producers in the region has been increasing over the last few years. According to the Afar Regional Pastoral and Agriculture Development Bureau (ARPADB), the current number of charcoal producers in the region is estimated at 20,000 (most of them being highlanders), with an estimated production volume of 2,000,000 sacks of charcoal per month.

In Afar, charcoal business has increased in the last three to five years. Commonly, from 2000 to 3000 sacks of charcoal are estimated to enter the regional capital (Semera) of Afar daily, the main means of transport being ISUZUE medium duty trucks. However, there is seasonal variation in supply, demand, and price of charcoal in the region. The production of charcoal seems to diminish during Kiremt season while the price is at its maximum. On the other hand, prices are lower due to abundant productions and lesser demand during the Bega (dry) season. For instance, producers can sale a sack of charcoal for birr 35–40 during Bega season, and over birr 50 during Kiremt season. On average, the retail price (per 50 kg sack) of charcoal in the marketing chain from producers through distributors to consumers (Semera) is for birr 85, 110 and 140, respectively.

While most of the charcoal produced in the region ends up in the national markets, there are some illegal exports, mainly to Djibouti, Kuwait and Saudi Arabia. Initially, there were licensed cooperatives involved in charcoal production in the region. There is no national tax rate for charcoal. But, there were some checkpoints which regulate and collect tax from the business: a 250–280 sacks (50 kg) capacity ISUZU medium duty truck was taxed about birr 300.

Though not successful, FARM Africa tried to facilitate charcoal production through, for example, the introduction of improved charcoaling metal kilns and giving some extension services related to controlling the spread of the invasive species–P. juliflora. They placed checkpoints to regulate charcoal making and the activities of the established cooperatives, with the support of FARM Africa with the aim of promoting charcoal making from P. juliflora. This has not, however, succeeded in controlling the spread of the invasive species (ARPADB).

Beyond its impact on the health of the producers and consumers, and the environment, charcoal production is becoming a source of conflict among clans in the region. However, there is a growing market access and demand at local,

national and international markets. Moreover, the fast growing invasive species–
P. juliflora is providing continuous supply to the production. There are also
experience sharing among producers (from the highlanders), and at the same
time new improved charcoal making technologies are being introduced (e.g. by
FARM Africa). These all issues can be treated as opportunities for the charcoal
business in the region.

4.5. Bahir Dar (Amhara Regional State)

The Amhara Region is located in the northwestern part of the country and has a
land area of about 170,752 square kilometers (Woreta, 2007). The total
population estimate for the region for mid-2008 is 20,136,000 (CSA, 2008), with
a fifty-fifty split between the sexes. Of these, only 12% are urban residents. In
the region, there are a total of 169 urban centers with a population of 2,000 or
more. With a population of about 213,673, Gondar is the most populous town,
followed by Dessie and Bahir Dar. Two-thirds of the urban centers in Amhara
region have a population of less than 10,000, and more than a third have a
population of less than 5,000 (Aynalem, 2008). Majority (about 52%) of the
region's land area is under cultivation, while the forest resource base covers
about 5.91% of the total regional area. The vegetation resource is dominated by
the dry-woodlands (about 4.2%). High forests account only 0.48% and plantation
forests cover 1.23% (Woreta, 2007).

Alike most regions in Ethiopia, the charcoal business in Amhara region
remained informal. Reports (e.g. Abebe, 2004) indicate that there are hardly
licensed/legal charcoal producers in the region; hence it is difficult to quantify
the number of producers. However, a considerable number of the rural poor are
assumed to be involved in charcoal making activities to support their livelihoods.
Most households in Bahir Dar, the regional capital, still rely on wood-fuel
(mainly charcoal) to satisfy their household energy needs. Charcoal to the town
is mainly sourced from the nearby districts as far away as, among others, Merawi
and Zegie. The means of charcoal transport to the town is mainly through trucks
and sometimes through donkey loads. Most charcoal producers in these districts
are farmers and the landless youth. The farmers use charcoal making as way of
supplementing their livelihood, while the landless farmers may entirely depend
on it. While charcoal making is a seasonal activity for farmers, it is a year round
occupation for the landless.

According to Rural Energy Resources Development Agency of Amhara Regional State (2011), about 86% of charcoal producers partly depend on charcoal business, as additional source of income; while the rest, 14% entirely depend on it as means of livelihood. Charcoal making is mainly of men's work; while some women involve in charcoal production to help men through supplying water or transporting the wood used for charcoal making. According to charcoal producers, the main reason they joined the charcoal business was due to frequent drought and low agricultural productivity and in some cases the relatively good income obtained from the charcoal making as an off-farm activity.

Charcoal mainly comes from the wood and shrub lands found in the arid and semi-arid agro-ecology of the region (Abebe, 2004; Rural Energy Resources Development Agency, 2011). But, some producers use their private plantations (mainly eucalyptus) for charcoal making although its quality is considered to be low due to its sparks. Tree species used for charcoal making in the region vary with different agro-ecologies, however, due to the increment in the number of producers, and natural forest degradation, charcoal is being produced from every available plant species regardless of its quality. For instance, species such as Ge'rawa (Vernonia amygdalina) and Be'sana (Croton macrostachyus), which were considered less convenient for fuel-wood, are currently main tree species used for charcoal making. Some of the most preferred charcoal species include: Abalo (Terminalia brownie), Wanza (Cordia africana), Sesa (Albiza gummifera), and Girar (Acacia sp).

The assessment showed that the overall trend in production, consumption and price of charcoal is found to be increasing in the region in the past 3–5 years. As is with other regions of the country, the increment in charcoal demand is attributed to the price escalation of other modern energy alternatives (electricity, kerosene, LPG etc.) on one hand; and its easy access and affordability on the other hand.

Expropriation of illegally produced charcoal has not been effective to deter production and trading. This is mainly because the economic need to produce charcoal is growing among the rural poor, while the demand for it in the urban areas is increasing. There are, however, some emerging opportunities, such as introduction of improved charcoal making technologies and improved stoves in some parts of the region. The regional Rural Energy Resources Development

Agency, in collaboration with Agricultural and Rural Development offices, has started to create awareness among the public and identified some improved charcoal making technologies (e.g. Casamance, portable metal kiln) suitable to the region.

4.6. Adama (Oromiya Regional State)

Oromiya National Regional State is the largest regional state in Ethiopia in terms of its area and population size. It has an estimated area of 353, 690 km2 (accounting for about 32% of the total area of the country). Based on the 2007 census of the CSA, the region has an estimated total population of 27,158,471 (accounting for about 36.7% of the country's population); with nearly equal proportion of men and women (CSA, 2008). With an average annual growth rate of 2.9%, the region's population growth is higher than the national average (which is 2.6%). Of its total population, nearly 3.4 million are estimated to be urban residents. With about 12.2% of its population living in urban areas, the urbanization rate of the region is slightly below the national average (16%). Administratively, the region is divided into 18 zones, 304 woredas (out of which 39 are towns structured with the level of woredas and 265 rural woredas). Twenty three woredas have a population of 200,000 or more, and 31 woredas have a population of less than 100,000 (CSA, 2008). The 2007 Census of the CSA estimated the total population of the urban residents in Adama at 220,212, which showed an increase of 72.25% over the population recorded in the 1994 census. With about 69% of its total area being under vegetation cover, the region represents the largest vegetation cover in the country. Afro-Alpine and Sub-Afro Alpine, High forest, Woodland, Riverine, Grassland, Plantations, and Bush and Shrublands are the major types of vegetation forms in the region (BoFED, 2008).

The available energy sources in the region includes biomass fuels (wood-based and animal dung), kerosene, electricity, and to some extent solar and bio-gas. Charcoal is particularly used most frequently among the urban residents. According to the discussion with charcoal distributors, retailers and consumers, and government bodies (forestry department) of Adama, one can find charcoal being used in almost every household either as supplementary or main energy source. In some cases, especially among the poor, animal dung and firewood are also extensively used. Similar to other areas, the charcoal industry in the region is dominated by men; except that women can, in some cases, participate during the preparation of charcoal making sites and/or supplying raw materials.

Most charcoal makers in the region are farmers engaged in charcoal making as an off-farm activity (during less-work load season) to supplement their livelihoods. It is reported that charcoal is produced informally from the natural woodlands of the region on an open-access basis; but to a lesser extent charcoal also comes from private eucalyptus woodlots. The most preferred species widely used for charcoal making in the region are Acacia species (e.g. Acacia albida and A. abyssinica) followed by Prosopis juliflora. Most of the charcoal produced in the region ends up in the domestic market. But, the assessment revealed that charcoal produced in the region is also exported illegally enclosed in containers of heavy load trucks to Djibouti.

The charcoal industry in Oromiya region is mostly informal and/or illegal and not properly regulated: there are hardly any legal charcoal cooperatives, no effective checkpoints and no taxation systems developed. As a result of the lack of proper attention on the sector, assessments made on charcoal issues, be it by GOs or NGOs is negligible. Thus, the volume of charcoal produced, and consumed in the major urban areas of the region, including Adama, is not properly known.

The charcoal supply to Adama comes mainly from the Acacia-dominated woodlands of the Rift Valley and Afar region; the main means of charcoal transport being by car (mainly ISUZU medium duty trucks), pack animals (mainly donkeys and camels). Even though the amount of charcoal entering the town is unknown, some informants during the field assessment indicated that a given charcoal distributor (depot owner) can receive about 200 sacks of charcoal a week. The price of charcoal through its marketing chains varies considerably. Based on the current charcoal market and according to interviewees (distributors, retailers and consumers), the average retail price (per 50 kg sack) of charcoal in Adama, the marketing channel from producers through distributors to retailers is estimated at birr 80, 100 and 125, respectively.

Obviously, there are seasonal variations in the supply, demand and price of charcoal in this region as well. While the demand and price of charcoal shows a relative increment, the production decreases during kiremet season since most farmers are engaged in agricultural activities. However, the assessment showed that the overall trend in the supply, demand, and price of charcoal has been increasing in the last three to five years.

4.7. Hawassa and Arba-Minch (SNNPR)

Located in southern and southwestern Ethiopia and covering an estimated total area of 114, 781 square kilometers, the Southern Nations, Nationalities and Peoples Regional State (SNNPRS) is one of the four largest regions of Ethiopia, accounting for more than 10% of the country's land area. The SNNP agro-ecology ranges from arid to semi desert in the Omo river lowlands inhabited by transhumant pastoralists, to montane forests with high rainfall inhabited by bush fallowing agriculturists (WBISPP, 2001). The region's population is growing at a faster rate of about 3%. According to the report by CSA (CSA, 2008), with over 15 million inhabitants in 2007, the SNNPR is the third largest populated regional state next to Oromiya and Amhara. With only about 10% of its population living in urban areas, the region is among the least urbanized regions of the country compared to others (e.g. Diredawa and Harari) where more than half of their populations are urban residents. The capital of the region, Hawassa (with 157,139 residents), and Arba Minch (with 74,879 residents) are among the most populous towns in the region (CSA, 2008). With an estimated forest area of 1.3 million ha (10.3% of the national woody biomass cover), SNNPR is the fourth region with highest forest cover in the country next to Oromiya, Benishangul-Gumuz and Gambela. It has a relatively better endowment of forest resources including the nation's renowned high tropical forests of south-west (Melessaw and Hilawe, 2011).

Hawassa:

According to charcoal production and marketing assessment made in September 2012, the main source of household energy in most towns in SNNPR region, including its capital – Hawassa, is composed of firewood and charcoal, followed by electricity and kerosene. A similar study by Melessaw and Hilawe (2011) confirmed that firewood (including branches and twigs) and charcoal are the two most important fuels commonly used by urban households in SNNPR.

The charcoal supplied to Hawassa comes mainly from Blatie district in Wolayta zone. Though most of the share of producing charcoal in this district is by men whose age is between 25 and 40 years, women are found to participate in charcoal production at least through collecting raw materials to facilitate the production process. Charcoal production is not the main stay for most of the producers in Bilatie, rather it is supportive business meant for additional income. Producers rely on the prevailing common forests/woodlands to produce charcoal on an open - access basis. Many of the Acacia species, Eucalyptus, and Croton macrostachyus are among the commonly used charcoal species in the district.

Different types of trucks (mainly Isuzu medium duty trucks) and donkey carts are the major means of charcoal transportation to the Hawassa town. The average selling price of a sack of charcoal in the chain of charcoal supply route into the town is estimated at birr 75, 130 and 165 for producers, distributors and retailers, respectively. While seasonal variations are acknowledged (e.g. production is reduced considerably during rainy season), the overall trend in supply, price and consumption of charcoal in Hawassa town is found to be increasing in the last three to five years.

Even though there were some attempts by agricultural and rural development offices to monitor charcoal business in the town and its vicinity, currently, there are neither functional extension services nor checkpoints to control illegally produced charcoal entering the city. While the growing demand for charcoal among the growing urban population can be considered as opportunities for charcoal business in Hawassa town and its main charcoal supply areas, the production process and the lack of preferred species in connection to the escalating forest depletion are mentioned as main challenges/threats to the business.

Arba-Minch:

Firewood and charcoal were mentioned as the main household energy sources for the urban residents in Arba-Minch town. Charcoal to Arba-Minch is supplied from its vicinity (around 10 to 20 km) on foot, and up to a distance of 100km from the town by cars. The main source of charcoal for the city is the nearby state forest, private land and investment areas; the majority being accessed freely. Acacia species are the most preferred species for charcoal making. Men in the age range of 18 to 45 years are involved in the production, and distribution in town. Women's role can be mentioned in connection to charcoal transportation into market areas.

Even though seasonal fluctuations in production, marketing and consumption of charcoal are evident, the overall trend in charcoal production in the vicinity of

Arba-Minch seems to be declining, while the demand and price of charcoal is increasing. This could be attributed to urban expansion and continued depletion in forest resources which in turn brought about shortage of supply of raw materials for charcoal making.

Just like most areas in the country, there is no separate responsible body which monitors or provides extension services to the charcoal sector around Arba-minch. It is merely illegal business and hence is posing severe threats to the existing forest patches and woodlands. According to the discussion made with concerned bodies (Agricultural offices) in the district, the unwise exploitation and conversion of woodlands by investors (meant for agricultural investment) is even worsening the situation: they involve in illegal production and marketing of charcoal.

4.8. Harshin (Somali Regional State)

The Somali Regional state is located in south-eastern part of Ethiopia, and is one of the nine states with a total area of 350,000km2. According to the 2007 Census, the Somali Region has a total population of 4,439,147, consisting of 2,468,784 (55.6%) men and 1,970,363 (44.4%) women with a population growth rate of 2.6%, which is similar with the national average (CSA, 2008). The census has estimated urban inhabitants in the region at 621,210 or 14% of the population. The region has a low population density standing at about 15 persons per km2. It is estimated that more than 85% of the population in the region are pastoralists. The vegetation of Somali region is endowed with Acacia, Boswellia and Commiphora (ABC) species that are sources of important forest products like gum, incense and myrrh which are sold both at domestic and foreign markets.

As Sead's (2007) report confirms following the ban of livestock export to the Middle East in 1998 and due to the intensified uncertainties from recurrent drought, commercial level charcoal production and trade came into view among the pastoralists in Somali region. In an informal interview with two long time resident of Jigjiga, the capital of Somali Regional State, it was learnt that charcoal making was not among the traditional work-list of pastoralists. The Somalis were prohibited from cutting trees and offenders were fined with livestock depending on the extent of the damage and if they had previous record (PFE, IIRR and DF, 2010).

Sead (2007), who studied the changing land tenure condition in Somali region, reported that charcoal production started in the region in small scale some 50 years ago. By then, the charcoal industry was not an important source of livelihood, but it was simply considered as a last resort for poor households. It was only those poor households with few cattle or without any other livelihood support system who were engaged in charcoal production. These days, however, the main actors in the charcoal industry are "wealthy businessmen, often from Somaliland, who organize and finance mass production through the mobilization of local communities" (Sead, 2007).

Owing to the increment in charcoal demand and the loosely organized control on the sector from both sides of the border, many people have been involved in the production and illegal export of charcoal into Somaliland. The easternmost district of Jigjiga zone, Harshin town is an important center where a thriving charcoal trade with the neighboring Somaliland is evident. According to Hagmann (2006), in 2005 Oxfam GB calculated around 63,000 sacks of charcoal harvested from Harshin district alone and transported across the border to Hargeissa on a monthly basis utilizing 27,300 trees.

The report noted that the charcoal trade is conducted with little control at the borders; as a result, the charcoal business has attracted different groups[13] of stakeholders from various backgrounds, including khat sellers, soldiers, professional business men, haulage companies, women and youth. For some groups (the poor pastoralists) it is a livelihood necessity, whereas for others it is for asset saving and wealth accumulation (Sead, 2007). The author furthermore argues that the social division being created between the old and the young generation within the pastoralist community is turning the latter away from the old way of life. In the majority of cases it is the young drop-outs from school addicted to Khat and cigarettes who are engaged in charcoal production (Sead, 2007).

---

[13] Main groups involved in charcoal business around Harshin: (i) poor pastoralists who continue to be dependent on such activities due to their limited livestock resources; (ii) pastoralists with adequate livestock assets to meet subsistence but try to generate additional income from the freely available communal resources as an income buffer and guarantee against shocks; (iii) women who own small retail businesses and sell charcoal in the towns and villages; and (iv) young people aged between 17 and 22 who have abandoned livestock rearing to live in the towns and villages (source: Sead Oumer, 2007).

Charcoal production in the region is conducted in an inefficient manner from, mainly, acacia trees. As is the case with many areas in Ethiopia, here also trees are accessed free of charge from state-owned woodlands resulting, according to Sead (2007) and Fikre et al. (2010), in permanent rangeland depletion and increased soil erosion from wind and runoff. The report by CHF International (2006) also confirms that due to intensive charcoal production, particularly in the last two decades, areas around Jigjiga suffer from the effects of deforestation. Decline in the important browsing tree and scrub species in the region is greatly affecting the pastoralist way of life (Jama and Walker, 1998). The severe depletion of forest resources in the border areas between Ethiopia (Harshin district) and Somaliland has induced conflict in the borders on the sale and utilization of trees from both inside land enclosures as well as from the remaining communal lands. Conflicts often arise among the traditional resource users (pastoralists) and the emerging stakeholders (charcoal producers, district authorities, farmers, etc.) (Sead, 2007). Moreover, the depletion in rangeland and tree resources pushed individual households to create private enclosures (PEF, IIRR and DF, 2010).

4.9. Dire Dawa City

Dire Dawa is Ethiopia's second largest city, located in the Great Rift Valley of Ethiopia. According to the 2007 census conducted by the CSA of Ethiopia, Dire Dawa has a total population of 341,834 (including the surrounding rural countryside), with a balanced gender composition (171, 461 are men and 170,461 women). About 68.23% of the population is urban inhabitants. Its population has, generally, increased by about 30% compared to the 1995 census, and is expected to grow by 50% more by the year 2015 (CSA, 2008). The annual rate of population growth for Dire Dawa city is almost the same as the national rate, which is about 2.6 percent (CSA, 2008). Most inhabitants in the vicinity of Dire-Dawa depend on dryland crop cultivation, livestock rearing, mainly camel and charcoal business for their livelihood. The vegetation formations in the vicinity of Dire-Dawa city is dominantly covered by shrub and woodlands dominated by acacia species. In the last two decades, the invasive tree species, Prosopis juliflora has taken over large areas in the vicinity of the city.

Firewood and charcoal, followed by kerosene and electricity are the main household energy sources for the urban households in Dire Dawa. Charcoal supplied to the city comes mainly from the acacia dominated dry free-access woodlands from localities in the vicinity of the city within a radius of about 50km. There are, however, some trends where charcoal is transported from the Afar region. Acacia species (locally known as t'edecha) are the most preferred charcoal species. But, nowadays, due to scarcity of supply, many other indigenous tree species and some other exotic species including Prosopis juliflora are being used to produce charcoal. Alike most of the charcoal making technologies in different regions of the country, charcoal making techniques in Dire Dawa are primarily based on the traditional kilning methods (earth pit or mound kilns). The field assessment showed that an estimated volume of 1200 sacks of charcoal is produced daily from the surrounding areas.

People involved in charcoal production and marketing in the region are men aged 18 to 35 years. Like other regions, charcoal business in Dire Dawa is dominated by illegal actors. Thus, it is difficult to quantify the number of producers and actors in the business as well; it varies considerably. According to information received from charcoal brokers and distributors, camels, followed by donkeys and cars serve as the main means of charcoal transportation into the city. It is estimated that 80 to 100 camels, and 100-120 donkeys, with an estimated amount of 800 sacks of charcoal enter into the city daily.

Generally, the trends in production, marketing and consumption of charcoal in the city showed an increment in the last 3–5 years. In the charcoal marketing supply in the region, brokers serve great role as messengers among the various actors from producers to consumers. The estimated price of charcoal (per 50 kg sack) in the marketing chain from producers through distributors to retailers is estimated at Birr 85, 105, and 90, respectively.

Generally, charcoal production is confined to illegal actors. There are no legal cooperatives and no extension services in regard to charcoal production and marketing in the city. While appreciating opportunities associated with increasing demand and introduction of modern charring processes, environmental degradation and the scarcity in raw materials have been mentioned as major threats to charcoal business in the region.

5. Charcoal Impacts

5.1. Charcoal Impact Assessment

Although there is little information concerning the impact of charcoal making on the environment and human health in Ethiopia, related studies elsewhere showed a direct impact of charcoal on forests, soils, climate as well as human health. A study by Kammen and Law (2005) on charcoal impact showed that regardless of the cooking advantages of charcoal and its high placement on the biomass cooking ladder, it may be far more damaging to the environment than the less preferable biomass fuels, biomass residues and firewood. When charcoal sources (trees) are unmanaged, and the utilization is unregulated or free, charcoal production causes lasting deforestation and environmental degradation. Such practice has also twisted the views of people in the charcoal industry to look at the trade negatively (UNDP/UNDESA/WEC, 2000).

In the following pages, charcoal impacts on forests/woodlands, climate change, soils, and human health is reviewed. As the main sources of charcoal in Ethiopia are its natural forests and woodlands, relatively extended review is provided on these resources together with the charcoal impact on climate and soils.

5.2. Forest Resources of Ethiopia and the Charcoal Impact

5.2.1. Forest and Woodland Resources

In Ethiopia, there is lack of reliable information on the forest resource base; estimates remained inconsistent. This has been reported as one of major impediments to planning and implementing sustainable forest management interventions (EPA, 2008). The Ethiopian forest resource base has entertained various estimates at different times, which vary considerably: some of which include the report by the Ethiopian Forestry Action Plan (EFAP, 1994), the Forest Resources Assessment (FRA, 2000), and the Woody Biomass Inventory and Strategic planning Project (WBISPP, 2004).

According to the WBISPP in 2004, the total area of high forest of the country was estimated at 4.07 million ha (about 3.56% of the total area of the country). Woodlands and shrubland types are the other most widespread vegetation resources of the country. An estimated area of 29.24 million ha (about 25.5% of

63

the total land area) is covered by woodlands, and about 26.4 million ha (23.1% of the total area) by shrublands (WBISPP, 2004).

Forest plantation, mainly composed of exotic species, is the other vegetation cover in the country. According to Million (2011) the area of plantation in Ethiopia is estimated at 972,000 ha. Excepting for the 190,000 ha classified as commercial used for the production of sawn wood, the remaining plantations are used to produce fuel and construction woods. However, it is common to observe very low productivity in most plantations due to poor management interventions (EPA, 2008).

The distribution of the vegetation resources in Ethiopia is uneven; with considerable variations in extent and type of vegetation covers within each region in the country. Most of the forest cover of the country is confined in some regions: Oromiya, Benishangul-Gumuz, Gambela, SNNPR and Amhara. According to WBISPP (2004), about 95% of the total high forest is located in three regions namely, Oromiya (63%), SNNPR (19%) and Gambela (13%) regional states. On the other hand, Melessaw and Hilaw (2011) compiled current FAO reports and put the highest woody biomass cover of the country to be distributed in Oromiya, Amhara, Benishangul-Gumuz, SNNPR and Gambela regional states (Figure 21).

Figure 21: Percentage Distribution of National Woody Biomass Cover by Regions (Ethiopia)

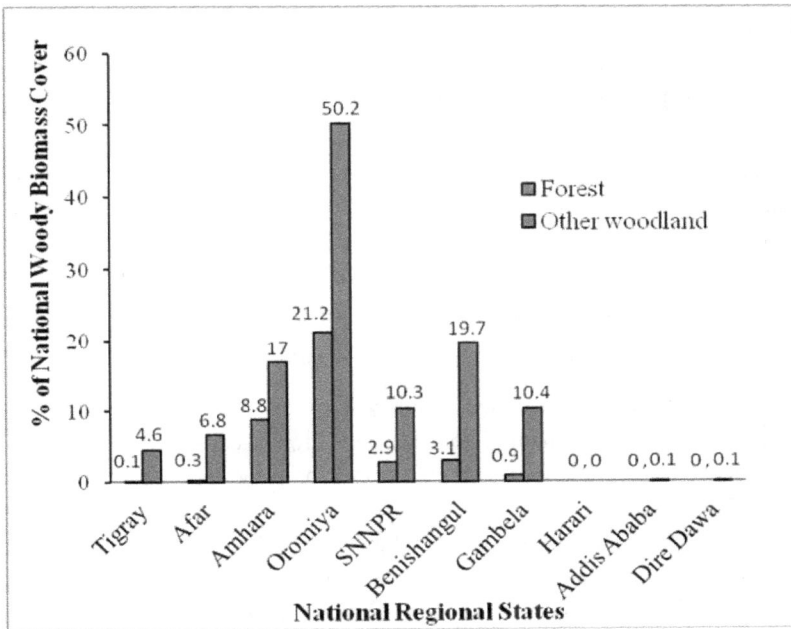

SOURCE: Adopted from Melessaw and Hilawe (2011).

The dry-forests, which accounts for 55 – 60%, are Ethiopia's largest vegetation resources (Mulugeta and Habtemariam, 2011). Even though variations exist depending on geographical location, vegetation formation, species type and the degree of disturbances inflicted, the regeneration profile of most species in the dry-forests of Ethiopia are generally poor due to an open-access nature of exploitation. The remnant vegetation resources of the country are receiving severe upheavals from the ever-increasing human-induced and natural stresses (Mulugeta and Habtemariam, 2011); thus they are being depleted at an alarming rate. For instance, according to the 2010 FAO report, Ethiopia lost over 2 million ha of its forests between 1990 and 2005, with an estimated mean annual loss of 140,000 ha. On the other hand, Yigard (2002) put the estimate of the annual forest cover loss of the country between 150,000 and 200,000 ha.

Expansion of commercial agriculture, state-sponsored settlements and encroachment, over-grazing by domestic animals, excessive exploitation for fire-wood and charcoal, destructive and undue tapping of gum and resin has resulted in large scale deforestation in the dry woodlands of Ethiopia. Various biophysical and socioeconomic studies (e.g. Abeje, 2002; Bongers and Tennigkeit, 2010; Abraham et al., 2010; Tefera, 2011; Teshale, 2011, Mulugeta and Habtemariam, 2011) (and other studies from Somali, Borena, South Omo, Afar regions) show that the woodlands of Ethiopia are under huge pressure from a range of factors. Some of the above studies concluded that if such type of exploitation continues, the dry forest resources would enter a state of total depletion with far-reaching consequences to the more fragile dry-land eco-system and community's livelihood in not distant future.

5.2.2. The Charcoal Impact on Forests and Woodlands

The bulk of fuel-wood consumed in many large towns and cities in the world comes from the tropical natural forests. These forest ecosystems are the most diverse ecosystems in the world ranging from closed moist (rain) forests to open woodlands and scrublands. They are found in at least 114 countries in the world and cover about 1915 million ha (Chidumayo, 2011). In almost all countries where charcoal is produced, there have been reports highlighting concerns about deforestation[14] and forest degradation[15] linked to charcoal production. Nevertheless, since fuel-wood is collected primarily for subsistence use and the charcoal trade is informal, and in some countries illegal, statistics are generally very poor. The FAO states that 'informally or illegally removed wood, especially fuel-wood, is not usually recorded, so the actual amount of wood removals is undoubtedly higher' (FAO, 2006).

Wood charcoal meets an overwhelming proportion of energy needs in developing countries, and will continue to be a significant fuel to millions of households for years to come. It is preferred fuel among urban households; thus, its production will not diminish in the near future. Consequently, the pressure on forests from charcoal is expected to grow in the 21st century as Africa is going to be more urbanized: "migrants switch from firewood to charcoal as they adapt

---

[14] Deforestation is the complete loss/clearance of forest cover

[15] Forest degradation refers to less obvious changes in the woody canopy cover; reduced productivity

the cooking habits of the urban environment" (Girard, 2002). Recently, the shift from firewood to charcoal, especially in Africa has raised concerns among environmentalists and those responsible for forest development and management (Girard, 2002).

In Ethiopia, although there is lack of reliable information concerning to the impact of charcoal on forest and woodland resources, fuel-wood extraction (both firewood and charcoal) is mentioned among the prime factors causing natural vegetation depletion, thereby worsening ecosystem degradation in the country. Similar to the cases with other developing countries, the charcoal industry in Ethiopia has been viewed negatively as it is often associated with the escalated rate of deforestation and degradation, slow and unsustainable growth of trees, inefficient use of wood, environmental pollution and poor working conditions of charcoal producers (Yisehak and Duraisamy, 2008; Shiferaw et al., 2010).

With the increasing population growth and rapid urbanization in Ethiopia, the demand for charcoal as energy source is increasing. Currently, charcoal production is commonplace in the arid, semi-arid and dry sub-humid parts of the country. The dry-woodlands (especially the Acacia-dominated woodlands) of the country, which have been important sources of charcoal, are under excessive deforestation and degradation, with the stock of the preferred species, like that of acacia, being depleted. The case is even severe in the Rift Valley areas, especially in the areas lying towards the lower Awash Valley, which supply charcoal to nearby towns and to Addis Ababa (EPA, 2008).

The prevailing charcoal production systems in Ethiopia are unsustainable; the raw materials for charcoal come from free sources, production technology (which uses the traditional charcoal kilns) is highly inefficient and there are little incentives for investment in the charcoal industry, particularly in the establishment of plantations for charcoal production. This all has contributed to the dismal pictures on the industry.

First of all, it was observed that the bulk of charcoal produced in Ethiopia is, very often, harvested illegally from the natural forests and woodlands on an open – access basis. There has been neither a concerned body nor regulations outlined regarding the charcoal sector in the country. As charcoal is habitually produced in "free" sources, it is hardly ever associated with sustainable forestry practices (Kammen and Lew, 2005). Other studies elsewhere (e.g. UNDP, 2000; World Bank, 2009; Chidumayo, 2011) have also confirmed that the unregulated, free or very cheap sources of trees and shrubs for charcoal production cause endless deforestation and environmental degradation, which in turn has negative impacts on the quality and quantity of various ecosystem services.

Mooney (1954) who visited the acacia woodlands in the Rift Valley found what he called "reckless cutting" of trees for fuel-wood. The area was heavily grazed and there was little tree regeneration to be observed, and as all trees were cut up to half a meter from the ground, a large amount of wood was wasted. He also (1957) reported that between 1954 and 1957 half of the acacia woodland between Modjo and Adamitulu (over 100 km apart) have disappeared.

The Franco-Ethiopian railway which ran from Djibouti to Addis Ababa through the semi-arid woodlands used acacia trees bordering the greater part of the railway as fuel for the wood burning locomotives (Logan, 1946). Russ (1944) reported that the uncontrolled cutting by the railway had left a large area denuded and permanently damaged. During the first 27 years of operation, the railway probably consumed a total of 3.2 million m3 of acacia wood on the basis of Logan's (1946) estimate of monthly consumption of 10,000 m3.

Secondly, in Ethiopia, charcoal is most commonly produced using the highly inefficient, traditional earth kilning technology. The traditional earth kilns as mentioned above are wasteful and requires larger quantities of wood to produce small amounts of charcoal. Yisehak and Duraisamy (2008) estimated the wood-to-charcoal conversion ratio in Ethiopian traditional earth kilns to be between 10 and 15%; while Yigard (2002) put it between 8 and 12%. Similar situations are reported in other African countries. For example, using the earth mound kiln, about 12% efficiency is normal in Zambia (Kalumiana and Shakachite, 2003), 11-15% in Tanzania (Ngerageza, 2003), and 9-12% in Kenya (Theuri, 2003). Moreover, according to the survey results of the Program for the Promotion of Household and Alternatives Energy sources in the Sahelian countries of Burkina Faso, Cape Verde, Mali, Niger, Senegal and Chad, the growth in charcoal consumption has accelerated deforestation in many of these areas due to often inefficient production processes requiring 5 to 8 kilograms of wood for every kilogram of charcoal (PREDAS, 2008).

Studies (e.g. Kammen and Lew, 2005) indicate that most part of the energy of the wood is lost in the process of its conversion to charcoal. This implies that charcoal users eventually use much more wood than direct firewood users. According to Rogers and Eliakimu (2008), about 18 trees of DBH (diameter at breast height) of 32 cm can be used to produce 1378 kg of charcoal. Extrapolations of this figure implies that about 10,023,126 trees of similar size are cut to satisfy the annual supply of charcoal to Addis Ababa city in 2012; thus, there is no doubt that this adds serious threats to the forest resources of the country.

Furthermore, Ethiopian charcoal producers rely on specific preferred species (e.g. Acacia species) to produce quality charcoal for long. In line with Girard (2002) and Naughton-Treves et al. (2007), it is observed that charcoal makers may be highly selective when choosing trees, going for species whose wood is thought to make better charcoal, such as hardwoods; or they may take every species and practice clear felling. Targeting specific species that produce quality charcoal, which grow naturally unmanaged (predominantly Acacia species), leads to the loss of the species, which, in turn, remain a big threat to the forest biodiversity (Mugo and Ong, 2006). The concentrated exploitation of a few species with a high density can adversely affect biodiversity. Some dense tree species have a high economic value, e.g. as a source of timber, unrecognized by the charcoal producers (NL Agency, 2010).

Currently, given the scarcity of preferred species for charcoal making, the bulk of wood charcoal produced from every plant species in the natural forests and woodlands. There is an increasing scarcity of fuel-wood in most parts of Ethiopia owing to the alarming rate of deforestation/degradation of the remnant forest and woodland resources. Even in some areas, mostly the drier parts of the country, people are forced to use roots of trees for fuel; this induces grave consequences from an ecological point of view. Unless improvements and affordable alternative energy sources are sought, traditional woodfuel production will continue to supply the millions of energy hungered households; the consequence being a complete clearance of the vegetation of the country followed by the irreversible energy crisis.

5.3. The charcoal Impact on Climate

Charcoal production phase is one of the most polluting stages of the charcoal chain, producing a number of gaseous pollutants during pyrolysis process that are likely to contribute to global warming (Kwaschik, 2008). Because charcoal is frequently produced using inefficient traditional technologies, the amount of particulate matters (PM) and noxious gases emitted during charcoal production is expected to be higher in most developing countries. Therefore, charcoal production affects global warming through the production and emission of various green house gases (GHGs), such as carbon dioxide (CO_2), carbon monoxide (CO) and methane (CH_4) (Brewer et al., 2010; Chidumayo, 2011).

According to Girard (2002), one ton of wood produces about 150 to 200 kg of charcoal and an emission of 0.365 ton of carbon into the atmosphere with a poorly managed carbonization technique. From this, one can extrapolate the amount of wood carbonized often through traditional and inefficient charcoal making technologies to meet annual charcoal demand in developing countries. For example, according to the charcoal inflow survey conducted in August 2012, about 1,471.575 tons of charcoal was supplied to Addis Ababa city in a day (an equivalent of 537,124.875 tons of charcoal per annum). The extrapolation of this imparted 2,685,624.375 – 3,580,832.5 tons of wood carbonized for the annual supply of charcoal to Addis Ababa, resulting in a release of 980,252.89 – 1,307,003.86 tons of carbon to the atmosphere. Compared to earlier estimates (e.g. Shiferaw et al., 2010), this value is larger implying an increased number of trees are being cut every year to satisfy the increasing demand of charcoal for Addis Ababa city. This all is escalating the deforestation rate, thereby contributing significantly to the greenhouse effect.

Some sources estimate that cooking with traditional biomass fuels contributes about 18% of current global GHG emissions when forest degradation and deforestation is included in the equation (SEI, 2008). If charcoal was sustainably produced it would be carbon neutral since this emitted carbon could be sequestered by trees that are planted. In this scenario, one ton of sustainable charcoal would make up for one ton of non-sustainable charcoal or nine tons of carbon dioxide (GEF, 2010).

Nevertheless, the burning charcoal, along with fire-wood, agricultural residues and animal dung, in traditional and inefficient stoves, produce high emissions of carbon monoxide, hydrocarbons and particulate matter pollutant to the atmosphere (Smith, 1993; Smith et al., 2000). Hydrocarbon emissions are highest from the burning of dung for fuel, while particulate emissions are highest from agricultural residues (OECD/IEA, 2006). The burning of biomass fuel, thus, not only affects human health (through indoor air pollution – IAP), but also adds a considerable burden to the environment (through the emission of GHGs). Charcoal is a more commercialized fuel and the nature of charcoal markets typically lead to greater woodland exploitation than firewood. This impacts the net GHG emissions resulting from charcoal production and can result in local environmental degradation (Ribot, 1993; Ellegård and Nordström, 2003).

Analysis of emissions from charcoal production in earth-mound kilns in several developing countries (Pennise et al., 2001) revealed that charcoal production is an extremely GHG intensive activity because it is essentially wood pyrolysis with the gaseous products vented to the atmosphere. Comparative studies showed that each meal cooked with charcoal has 2-10 times global warming than a meal cooked by a firewood, and 5-16 times the effect of the same meal with kerosene or LPG (depending on the gases used) (Bailis et al., 2004). Similarly, Pennise et al. (2001) found that the total emissions from charcoal production and use in Kenya, one of the largest consumers of charcoal in SSA, are equivalent to emissions from transport and industry in the country.

> The CRGE (2011) is a strategy developed by the Government of the Federal Democratic Republic of Ethiopia to build green economy, to protect the country from the adverse effects of climate change and help realize its development goals.
>
> Ethiopia's current contribution to the global increase in GHG emissions has been practically negligible – represents only 0.3% (around 150 Mt CO_2e) of global emissions. However, emissions from deforestation due to agricultural expansion and fuel-wood extraction remained the main sources of GHG emissions. Of the 150 Mt CO_2e in 2010, more than 85% of GHG emissions came from the agricultural and forestry sectors. Forestry emissions are driven by deforestation for agricultural land (50% of all forestry-related emissions) and forest degradation due to fuel-wood consumption (46%) as well as formal and informal logging (4%).
>
> While the forestry sector is significant contributor of GHG emissions, it also offers a high abatement potential that even surpasses the estimated increase in emissions by 2030. If Ethiopia were to pursue a conventional economic development path to achieve its ambition of reaching middle-income status by 2025, GHG emissions would more than double from 150 Mt CO_2e in 2010 to 400 Mt CO_2e in 2030. The country has, therefore, initiated the Climate-Resilient Green Economy (CRGE) initiative to protect the country from adverse effects of climate change and to build a green economy that will help realize its ambition of

reaching middle-income status before 2025. The country's green economy initiative offers GHG abatement potential of nearly 250 M and 25%t by 2030. Two sectors – agriculture and forestry – are given particular attention: they contribute around 45% to projected GHG emission levels under business-as-usual (BAU) assumptions, and together account for around 80% of the total abatement potential.

5.4. The Charcoal Impact on Soil

An overwhelming reliance on biomass fuel by households in developing countries is exacerbating the rate of deforestation and forest degradation. Besides, it induces severe impacts on catchment hydrology (more runoff, less water uptake, etc.) and serious repercussions for the ecological imbalance at large (Chidumayo, 2011).

In Ethiopia, the excessive deforestation, which led to the depletion of tree stock, caused what is known as the household energy crisis (Hawando, 1997). The increase in the cost of fuel-wood, thus, challenging the already staggering living condition. This crisis led to consumption shift towards animal dung and crop residue as household fuels. Although there is strong cultural preference in Ethiopia to use fuel-wood and charcoal for cooking, this preference had been affected by the scarcity of wood and hence, people started using dung and crop residue which accounted for over half of the total households' energy use (World Bank, 1984).

On the other hand, in spite of its overall negative effects on soil fertility associated with the heightened deforestation, charcoal production process is assumed to improve some soil properties, thereby soil fertility at its production site. Although further investigation is required to ascertain the long – term effects of charcoal production on soil fertility and crop yield, several studies (e.g. Glaser et al., 2002; Ogundele et al., 2011; Abebe and Endalkachew, 2011) revealed that charcoal production process serve to improve soil fertility by direct nutrient addition and retention. In line with this finding, a study in Southern Ethiopia (Abebe and Endalkachew, 2011) showed that while bulk density, water holding capacity and clay content seem to decrease, most soil properties; such as soil pH, organic C (carbon), total N (nitrogen), available P (phosphorus), EC (Electrical Conductivity), CEC (Cation Exchange Capacity), exchangeable K (potassium), Ca (calcium), Mg (magnesium) and Na (sodium) were significantly higher in charcoaling site soils than in a corresponding adjacent field. This could be attributed to the charcoal residues and charred biomass left on the kiln sites. Moreover, Oguntunde et al. (2004) reported an increment in both grain yield (by 91%) and biomass yield (by 44%) of maize on charcoal site soils as compared to adjacent farmland soils.

5.5. The Charcoal Impact on Human Health

At present little is known about the health status of African charcoal makers. In contrast, the health effects of biomass use have relatively been studied in some detail (Seidel, 2008). Charcoal production has a considerable health impact to producers during the carbonization process. While allowing wood to be carbonized at higher temperatures, it emits various poisonous gases, vapor and other volatile organic compounds; hence, the production process has a variety of health hazards to the producer as well (Bouros and Samiou, 2001). Some of the health risks, associated with charcoal production are summarized in Table 16.

Table 16: Potential health risks during charcoal production

Stage of Production	Potential Hazards	Health Risks
Cutting the wood	Accidents during cutting, heavy physical labor	Injuries
Compiling the wood	Heavy physical labor	Back problems
Covering the kiln	Low risk	Little or none
Igniting the kiln	Low risk	Little or none
Tending the kiln during carbonization	Breaking of the kiln surface	Burns, exposure to fumes and smoke
Opening the kiln, unloading charcoal	Noxious fumes and smoke, dust, tar, hot charcoal	Co – poisoning, irritation of eyes and respiratory tract, burns

SOURCE: Ellgård, 1993/2001.

Of the charcoal production risks, the risks from physical labor (during assembling of the logs and cutting of the wood) seem to outweigh the hazards. Exposure to noxious fumes could be limited as charcoal production is done in open; reduces the concentration of CO and other volatile organic compounds. However, without doubt, the most dangerous task is the opening of the kiln, where workers are exposed to a variety of health hazards (heat, tar, smoke). It has been reported, that CO-poisoning and even fatal accidents occur, if the kiln is not completely extinguished when opened (Ellgård, 1993/2001).

The smoke generated during carbonization is a complex mixture of liquid, solid and gaseous compounds (FAO, 1987). Many are noxious such as nitrogen and sulphur oxides, benzene, aldehydes, organic acids and polycyclic aromatic hydrocarbons (PAH), just to mention a few of them. Most of them irritate the respiratory tract and especially PAH are known to be carcinogenic, and exposure to wood smoke increases the risk of certain types of cancer of the upper respiratory tract and the oral cavity (Pintos et al. 1998). Measurements of suspended matter at kilns in Zambia have shown that concentrations are about fourfold of the level housewives are exposed to (Ellgård, 1993/2001), whereas CO-concentrations are the same. However, during charcoal production the health effects may be reduced by the fact that tending the kiln can be done in a rather short time, which reduces exposure time.

The results of interviews with charcoal producers (Gewanie and Bilatie) are in line with most of the health problems outlined above. The case was reported to be even worse among charcoal producers in Afar region, Ethiopia, where they face additional problems associated with attack from wild animals, such as lion and hyena. However, for African charcoal makers, the degree of exposure to health effects remains unknown; thus needs further endeavors.

Most African countries are heavily reliant on biomass fuel. Using dung, agricultural residues and fuel-wood in traditional stoves is inefficient due to incomplete combustion of fuel (Bailis et al., 2004; WHO, 2002). This produces large amounts of noxious gases and particles. When cooking is done inside, these pollutants lead to Indoor Air Pollution (IAP). The problem of IAP[16] is worse in developing countries where there are no separate living and cooking places. Women and children suffer most from IAP because they are traditionally responsible for cooking and other household tasks, which involve spending

---

[16] IAP mainly consists of two components: (i) CO - which is a color – and odorless but highly toxic gas; and (ii) PM - which consists of small, organic particles (also called soot) causing irritation and respiratory illnesses when inhaled. High levels of IAP cause a variety of health problems such as infections of the lower respiratory tract (Acute Respiratory Infection, ARI), chronic obstructive pulmonary disease and eye irritation (Source: WHO, 2006).

hours by the cooking fire exposed to smoke (OECD/IEA, 2006). Results from studies carried out in developing countries indicate that particulate concentrations from traditional biomass-using stoves are often ten or more times higher than the standards set by the United States Environmental Protection Agency (Albalak et al., 1999). Exposure to these high levels of pollution has been consistently associated with acute respiratory infections, the largest single-category cause of mortality worldwide (Smith et al., 2000).

Evidence also links exposure to biomass fuel combustion with chronic obstructive lung disease, tuberculosis, cataracts and adverse pregnancy outcomes (Albalak et al., 1999; Perez-Padilla et al., 1996; Mishra, Retherford and Smith, 1999; Mohan et al., 1989; Mavlankar, Trivedi and Gray, 1991). The World Health Organization (WHO) has estimated that as many as 2 million people in developing countries, the majority under five years of age, die prematurely every year from exposure to the combustion products of household solid fuels (Bruce et al., 2000; Albalak et al., 2001; WHO, 2002).

Since charcoal must usually be purchased, the introduction of energy-efficient charcoal stoves has been successful than the dissemination of firewood stoves. Burning charcoal emits less particle matter (PM), making charcoal a rather clean fuel than fire-wood. Although charcoal is worse than other fuels with respect to GHG emissions, it can lead to reductions in concentrations of pollutants like PM (Bailis et al., 2004). For instance, a household survey in Kenya found out that households using charcoal had significantly lower indoor concentrations of PM (about 88% lower than households using open wood fires), compared to those who use open wood burning (Ezzati et al., 2000; Ezzati and Kammen, 2001 and 2002; Bailis et al., 2004).

Whereas little PM is produced by combustion of charcoal under good cooking behavior, the high levels of CO emitted even by improved charcoal stoves induces severe health defects. Especially, charcoal burning in closed indoors may emit higher concentration of CO gas, which is highly poisonous leading to fatal death through disruption of free oxygen circulation in the body (Ellgård, 1993/2001).

In line with several studies (e.g. WHO, 2006; Kwaschik, 2008) elsewhere, respondents surveyed in Addis Ababa confirmed that indoor charcoal usage (in less ventilated rooms) is responsible for the production and inhalation of carbon monoxide (termed as the silent killer) which causes lung problems such as cancer and asthma. This amounts to 4% of the global burden of disease and leads in many cases to death. Women and children below 5 years of age are the most affected in developing countries (Kwaschik, 2008). Therefore, it should be noted that, besides using efficient charcoal stoves, behavioral changes, such as ventilation of the kitchen or lighting the charcoal outside, can greatly reduce IAP.

6. The Institutional Deficits in the Charcoal Industry: The Way Forward

6.1. Institutional Shortfalls

The complete absence of rules and regulations (institutions) to regulate the production, marketing, consumption, as well as impact of the charcoal industry in Ethiopia is mind-boggling in a country where charcoal serves millions of people as source of livelihood and energy, and its impact on the environment in particular needed close follow-up. In Ethiopia, charcoal is produced and marketed in a policy and legal (institutional) vacuum. And people cannot interact in institutional vacuum; even if they did, in most case it is costly. In the absence of regulatory institutions, uncertainty set-in this generates conflict in the ownership and use of resources, and lasting damage to the resource.

Regulated charcoal industry strongly influences the structural and legal settings in resource management and use; it is an essential input to understand the vital economic and environmental, as well as technological issues that could positively impact people's livelihood and the forest resources. The role the right institutions would have played in the sustainable construction of the charcoal industry in Ethiopia can hardly be overstated.

Sustainable charcoal production from wood is closely attached to how successful a country manages its forests in general, and in this case, its dry woodlands in particular. Nevertheless, in spite of significant contribution of the dry woodlands in Africa, few countries are making adequate investment in their management (Malimbwi et al., 2010). According to the same authors, there is a general lack of laws and regulations and/or their enforcement, absence of programs and political commitment to encourage the participation of the private sector and local communities in sustainable management of these resources.

The charcoal industry in Ethiopia has never enjoyed the benefit from policy direction and regulatory intervention or any technical support from concerned government agencies throughout the history of the modern state, with the exception of the 1980s[17].

---

[17] During the Italian occupation (1936-41) the *Milizia Forestale* started to experiment on the use of charcoal as a fuel for mechanically propelled vehicles and stationary internal

The most common intervention on the part of forest agencies for a long time remained to be the banning of what they considered illegal charcoal making without putting in place other livelihood alternatives. As learnt from the survey, the only relevant impact of outlawing charcoal production is more destructive tree resource utilization, inefficient way of charcoal making, loss of revenue by the state, and price increase.

One important policy direction that dominated the Imperial time was that forests were looked at as barrier to agricultural expansion. Compared to agricultural income, returns from forestry was too little at the time. If people wanted to own forests, it was for the soil under it. This notion very much dominated the forest policy direction in the country for long time across governments (Melaku, 2003). Forestry has its "heydays" during the Military Regime (1974-91) in Ethiopian history. The 1973/74 famine led to large rehabilitation work in different parts of the country. Soil conservation, re-forestation and afforestation works were given priority in order to rehabilitate degraded lands and increase land productivity. The government also initiated huge peri-urban fuel-wood plantation programs in Addis Ababa, Modjo, Gondar, and Bahir-Dar. Nevertheless, most forestry schemes which were initiated and established by state agencies resulted in excluding surrounding communities who lost their traditional grazing lands and even farm-lands for forest plantations. During the change of government in 1991, communities with grievances over lost lands overrun state-owned forests, destroyed forests and reclaimed the land. Survey made in 2002 showed an average of 49% state forest area reduction because of community action (Melaku, 2003).

In the last 45 years, the country has produced a number of forest laws and regulations. Although there is no one particular policy referring directly to charcoal, there are a number of other policies and laws that can be taken relevant to the charcoal industry. These include: the 1994 Energy Policy, the 2007 Forest Policy and Law, the 2005 Wildlife Policy and Law, the 2002 Environmental

---

combustion engines. No record has been found of their success in this field. During the time of the Military government an organization called Forestry and Wildlife Conservation and Development Authority (FaWCDA) was established. It used to give charcoal production and transport permission to organized charcoal producers, assign production area mainly around state farms, collect tax per sack (1. 5 birr per sack). There was also a fuel wood and charcoal marketing enterprise which was responsible for the marketing of charcoal.

Impact Assessment, the Conventions on Biodiversity and Desertification to which Ethiopia is a signatory, and other related legal documents and programs. Although the country received its first forest policy ever in 2007, the practical approach to develop the forestry sector is found to be inadequate taking into account the growing demand for wood products, including charcoal, in the country. The major impediment of natural resource and environmental protection in Ethiopia, however, has not primarily been lack of policies and laws and regulations, but their enforcement (Melaku, 2008).

Charcoal production is not by itself destructive. As Malimbwi et al. (2010) rightly commented, dry forest and woodland clearing for fire-wood and charcoal purposes may not have resulted in deforestation, provided that sustainable forest management is introduced in which production areas are identified, harvesting is done in accordance with management plan, improved charcoal production is introduced, protection of harvested areas from uncontrolled bush fires and overgrazing are restricted and, in some cases, tree planting initiated. But, as none of the above was fully implemented in Ethiopia, exploitation for fire-wood and charcoal have and still are taking their toll.

Probably for a long time to come, Ethiopia's major assets would remain to be its biological/natural resources. In recognition of the huge opportunities, the charcoal industry holds for society and the state as source of livelihood, energy and revenue, we would recommend the government to revisit the whole feature of the charcoal industry, and design a policy outlet where it can be organized on the basis of sustainable source of charcoal. As the charcoal industry is strongly attached to forest and woodland resources in Ethiopia, our suggestions focuses on sustainable management of these resources in general, although direct reference as well is made to the need to reorganize the business, re-organize the industry and develop charcoal sources.

6.2. The Way Forward

6.2.1. Charcoal as a Policy Agenda: Recognize and Legalize the Industry

Although the charcoal industry is generating and has the potential of creating additional jobs for thousands of wood producers, charcoal makers, transporters, distributors and retailers, and also provides energy for millions of urban dwellers and industries small businesses, it has yet to win the governments' recognition and policy direction. Lack of explicit policy direction prevented the charcoal industry benefiting from such input and extension services. For example, in Kenya, (where less charcoal is produced compared to Ethiopia) charcoal serves 82% of urban and 34% of rural households as source of energy. More importantly, the industry generates jobs for wood and charcoal producers, transporters and retailers numbering over 700,000. These in turn, according to the report, support the livelihood of over two million dependants. Kenya has therefore, developed tools that highlighted the issues in charcoal industry: tree planting, wood conversion to charcoal, transportation, trade and utilization (Gathui et al., 2011). Therefore, legalizing (decriminalizing) charcoal making, including it in the extension package, taxing the products and re-investing the income to develop forest and woodland resource and improving the technology of charcoal making are essential for Ethiopia.

Setting charcoal as an imperative policy agenda may start from opening a national dialogue. Through the national dialogue, all stakeholders (governmental agencies and environmental NGOs), higher education and research centers, civil associations and others will debate, with the aim of drawing a road-map (policy) for future undertakings, and creating a better awareness regarding the charcoal industry.

6.2.2. End Open-Access Situation

By contrast, sustainable forest management presupposes clear and secure long-term forest tenure ("property rights"). Since long time, the Ethiopian highland forests, and mainly the dry woodlands from which most of the charcoal is produced remained in an open access situation. Charcoal is produced from free sources, the woodlands – which is used by all and managed by none. It is free for the charcoal makers, but not for the public who bears all the environmental costs resulting from such activities. This is a crisis related to forest governance, particularly linked to the questions of who owns the Ethiopian forests, and how? It is a problem referring to a predicament in ownership. According to the institutional economist, Bromely (1991), most environmental problems are problems associated to property rights.

State-owned dry woodlands in the country are large tracts of land covering more than 28 million ha, i.e., nearly 25% of the land area of the country (WBISPP, 2004). It is difficult to put a thriving forest management system over such huge landmass by state agencies alone. It is, therefore, important to put forward an incentive system, including legal protection for dry-woodland property and call upon stakeholders to share management and even ownership responsibilities as well as the benefits. Partnership among stakeholders (state, community, private individuals) on the ownership and utilization of these resources could be a viable option. Governments' program to scale-up participatory forest management which is mainly contained on highland forests should be extended to include the dry woodlands.

6.2.3. Establish a Management System

Forest management is a branch of forestry that deals with both technical as well as social aspects of forests. The technical aspect deals with silviculture, i.e., establishment, growth, health, and quality of forests, while the social part deals with the policy/legal, administrative and economics of forests. A successful forest management consists of a means to keep a balance between consumption and conservation, i.e., sustaining the resource base while supporting livelihood and providing services. Chidumayo (2011) argue that the impact of charcoal on forest resources depend mainly on wood-stocking rate, tree cutting system, land tenure, etc. He notes that once cut, most tropical forest trees have the potential to regenerate, and suggest the importance of demarcating charcoal producing areas and put them under certain management system. The author reported the existence of such management system in some countries in West, Southern and East African countries.

In Ethiopia, there are no forests from which charcoal is being produced that are under any type of forest/woodland management. Therefore, Ethiopia needs to fill the existing knowledge gap in the woodland silviculture, the bio-physical characteristics of some of the important dryland tree species, and the management system that ensure their sustainability

6.2.4. Establish a Charcoal Agency and a Data Centre

The recognition of charcoal as a source of livelihood as well as source of energy shall lead to the creation of government agency at Federal and Regional levels responsible to organize all the affairs related to the charcoal industry. Important gaps that this survey revealed are absence of state agency directly responsible for charcoal production and marketing, and the extent of lack of information at the Federal or Regional level that refers to the industry.

One way of measuring state capacity is to look at its capability to make prudent policies and laws and enforce them for the general good of society. For this to happen, public policy decisions need to be based on relevant and adequately collected and analyzed information, according to Nutt et al. (2010). Ethiopia doesn't have a comprehensive data on various features of the charcoal industry: production source, type of technology, its contribution to livelihood and national economy, its place in household energy-mix, the market chain, potential of charcoal to generate employment and household energy, charcoal impact on the environment, etc. Without compelling evidence, it is nearly impossible to construct any prudent charcoal policy with realizable objectives. Therefore, the first task that should be given to a charcoal agency is the establishment of constantly updated data centre in order to assist policy makers and development agencies to make informed-decision about the industry.

6.2.5. Initiate Forest Plantations for Charcoal Production

As long as modern sources of energy remained expensive and inaccessible for the majority of the population, the importance of wood-based biomass energy will continue into the future (Kammen and Lew, 2005). In order to reduce deforestation and help meet demand for wood, plantations have to be promoted. World wide areas under plantation grew from 20 million ha in 1980 to 187 million ha in 2000 (FAO, 2003). Though plantation forest is expanding rapidly worldwide, there are marked regional differences. Asia and South America account for 91 percent of the 4.5 million hectares of annually planted areas globally. The area of forest plantation in Africa increased by less than 5% between 1990 and 2000, while in Asia it grew by about 20%, from 45 million ha to 60 million ha (FAO, 2003). In Africa, Kenya and Sudan regulate the charcoal trade; but, it is only Sudan that set-up large plantation for charcoal plantation in sparsely populated areas and under–used lands (Ibrahim, 2003). Unfortunately, apart from Sudan, the other African countries neither plant trees nor give land owners incentives to do so (Mugo and Ong, 2006). In Sudan, each year, 100,000 ha are planted with Acacia seyal and Acacia nilotica for charcoal. The wood is harvested for charcoal in rotations of 15 years. The Sudan Charcoal Association buys the wood from the state through tender. Charcoal traders pay taxes and other agreed fees to the government.

In Ethiopia, the government's conceptualization of forestry is restricted to farm-level trees (agro-forestry). Farm trees as important as they are for a household economy, are no substitute for big forest plantations. The latter's service extends beyond a simple household economy when their contribution to environmental services (soil, and water resource protection, climate change, and wildlife habitats) are taken in to account, leaving-out forests' huge potential for employment and the national economy. What is, however, encouraging at the present time is widespread engagement of many farmers' in tree-planting in small openings they have on their farm lands or homesteads.

There are no known plantation forests in Ethiopia that are meant for charcoal production. If currently charcoal is being produced from planted trees, it is from farmers woodlots found around homesteads. As important as they are, these homestead woodlots would not be able to cope-up with the growing charcoal demand. Large-scale forest plantations intended for charcoal and firewood production should be created. The policy response on the part of the government in such a case should then be able to develop a sustainably managed forest resources upon which the charcoal industry could be based and function with better benefit to society and with less impact on the environment. For this to happen, the government should take the first initiative and also call upon community and individual investors to invest in man-made forest plantation for the purpose of charcoal making and other related purposes. In such a venture, as investors need to wait extended time to benefit from their investment, the government should put together bundle of incentives for those who are willing to put their labor, money and knowledge in the creation of charcoal plantation.

6.2.6. Improve Charcoal Technology and Diversify its Sources

Charcoal making from wood has remained too traditional and wasteful; charcoal makers' skills are limited to the only and long-established technology they are very much accustomed to. The government and other concerned development agents/organizations need to support types of charcoal technology that reduces wastage and improve labour productivity. The dissemination of improved technologies will, thus, reduce the environmental burdens thereby enable the government to attain the goals set in the CRGE initiative of the country. As most of the households, particularly in rural areas, use highly energy-inefficient technologies, the improvement potential here is huge.

Although vital, wood is not the only source for charcoal making. Bamboo, Prosopis, agricultural residues such as cotton stalk, Khat stalk, and coffee husk, saw dust, have all shown promising initial results to satisfy demand in the short term and, if seriously promoted, to become possible alternates to wood charcoal (Yisehak and Duraisamy, 2008). There are some attempts in the country to produce charcoal from non-wood materials and wood by-products (agricultural residues, saw dust, coffee husks, etc.) with limited success. This technological initiative which remained underdeveloped needs to be institutionally and financially supported by the government and the concerned higher education and research organizations.

6.2.7. Develop Modern Energy Sources

Ethiopia's growing population and that of the urban residents, and expanding economy are obviously requiring huge amount of energy, both traditional and modern. Conveniently, the country is endowed with plenty of natural energy sources to meet this demand, primarily by exploiting its vast potential for hydro, geothermal, solar and wind power – all of which, according to EREDPC (2007), … "if adequately captured, the projected power supply could even exceed the growing domestic demand".

Despite the availability of huge energy resources, the current level of harnessing this energy is very low. This, to a certain extent, depicts the poor socio-economic situation in the country on the one side, and a low level of awareness about the potential and value of energy by most stakeholders on the other side (Ephrem, 2008). Energy development, if designed in line with the needs of agriculture, industry, transport and other related sectors, would highly accelerate the achievement of the development goals of the country. In order to confront this energy crisis and ensure sustainable development, the Ethiopian government planned to develop alternative renewable energy sources, such as wind, geothermal, solar, bio-fuels, together with energy efficiency measures (FDRE, 2012). This will be a key part of Ethiopia's energy mix and integrated with the country's new Climate Resilient Green Economy (CRGE) strategy, which has the ambitious objective to transform Ethiopia into a climate resilient green economy by 2025. Stress should, however, be given to the expansion and implementation of such energy development plans.

The development of such new renewable energies in Kenya, for example, has shown that diversification allowed for a stabilization of the power sector (Jacobs and Kiene, 2009). In line with the policy principles of the World Future Council (WFC), renewable energies can assure that natural resources can be used in a sustainable way, poverty can be eradicated, and human health can be improved by simultaneously protecting the ecosystem (Jacobs and Kiene, 2009). Therefore, the expansion and implementation of renewable energy technologies have a significant role in facilitating both social and economic development – it underpins economic activity, enhances productivity, and provides access to markets for trading purposes, thus, can play paramount role towards meeting the MDGs of the United Nations.

6.2.8. Education and Research

Sunderland (2011) noted that the dry forests and woodland resources received far less attention from research and development interventions than humid forest systems. As a result, the ecosystems are 'caught in a spiral of deforestation, fragmentation, degradation and desertification' (FAO, 2010). Although it is difficult to put all the problems in forestry to Sunderland's conclusion of lack of education and research, it certainly elucidate how the lack of education and training at university and technical and vocational level greatly contributed for the unfortunate situation of forestry including charcoal production. However, there are some attempts, for instance by Farm Africa, to orient the Afars about the benefits of turning prosopis in to charcoal and creation of charcoal cooperatives (FARM Africa, 2006).

This study showed that charcoal makers visited in Gewane and Bilatie were not organized and have never received training to enhance their efficiency and reduce wastage in charcoal production. However, charcoal makers, for example very well differentiate between trees. They know which trees makes good charcoal, which is smoky, etc. A study conducted by Tinsea et al. (2012) on local knowledge about fuel-wood and charcoal producing tree species around Awash National Park communities ranked Acacia tortilis as a tree that produces quality charcoal from among 11 acacia species.

Organizations such as research and education, Foresters Associations and similar civil societies, as well as relevant GO, E-NGOs, Think Thanks, etc. should be able to "invest" in policy research in the forestry sector. Such organizations can also establish a regular discussion forum to reflect on environmental/forest policies and laws of the country, at times inviting political leaders in order to contribute to prudent policies and laws of the country.

In conclusion, it has to be said that tree exploitation for charcoal making cannot alone explain the causes of forest resource depletion in Ethiopia. Furthermore, promulgating prudent policies alone cannot solve every problem in the charcoal industry. Any improvement in the industry is much dependent on the state's keenness to recognize, not only charcoal as an important source of energy and trade, but forest plantations, including the natural dry woodlands as viable economic sub-sectors by their own significance. It has also to be noted that the political as well as the humanitarian urgency that fall upon governments in Ethiopia to feed their population seems to have swayed them to give more attention to the production of food crops, in many cases, at the expense of forests and wildlife resources. This being the official line of thinking little could be done to change the grim situation in the forest/woodland sector. One has to recognize the unpleasant fact that governments can afford to lose forests as there is little immediate political risk in deforestation. Therefore, it takes strong effort and hard proof to convince the government that forests (natural and plantation forests) from which various forest products including charcoal are produced greatly contribute to food security even on an improved and sustainable ground compared to other land use system. One of the immediate themes of forestry research should be to show (in monetary terms) the economic benefits and the environmental significance of the sector to win the attention of policy makers and development agents.

References

Abebe Mohammed (2004). "በክልሉ ከሰል በሀጋዊና በተሻሻሉ ማክሰያ ቴክኖሎጂዎች የሚመረትበትን መንገድ የሚጠቁም ጥናት"። በአማራ ብሔራዊ ክልላዊ መንግስት፡ የገጠር ኢነርጂ ሀብት ልማት ማስፋፊያ ጽ/ቤት ፤ 1996, ባህር ዳር፡ ኢትዮጵያ። ("A Study to Indicate Ways of Legally Producing Charcoal Using Improved Technologies in The Region". Amhara Regional State Rural Energy Promotion Office. 2003/04. Bahir Dar, Ethiopia; in Amharic).

Abebe, N. and Endalkachew, K. (2011). "Effect of charcoal production on soil properties in Southern Ethiopia", Middle-East Journal of Scientific Research 9 (6): 807-813.

Abeje Eshete (2002). Regeneration status, soil seed banks and socio-economic importance of B. papyrifera in two woredas of North Gonder Zone, Northern Ethiopia. MSc Thesis, Swedish University of Agricultural Sciences, Skinnskatteberg, Sweden.

Abraham Abiyu, Bongers, F., Abeje Eshete, Kindiya G/Hiwot, Mengistie kindu, Mulugeta Limneih,Yitebitu Moges, Woldeselassie Ogbazghi and Sterck, F. J. (2010). in: Incense, woodlands, in Ethiopia and Eritrea: Regeneration Problems and Restoration Possibilities. Earthscan, Dunstan House, London.

Adam, J.C. (2008). "Improved and more environmentally friendly charcoal production system using a low-cost retort–kiln (Eco-charcoal)", Renewable Energy, 34: 1923–1925.

African Renewable Energy Access Program (AREAP) (2011). Wood-Based Biomass Energy Development for Sub-Saharan Africa: Issues and Approaches. Washington, DC. The International Bank for Reconstruction and Development.

Albalak, R., Bruce, N., McCracken, J., Smith, K.R. and Gallardo, T. (2001). "Indoor respirable particulate matter concentrations from an open fire, improved cookstove, and LPG/open fire combination in a rural Guatemalan community", Environmental Science and Technology, 35: 2650-2655.

Albalak, R., Frisancho, A.R. and Keeler, G.J. (1999). "Domestic biomass fuel combustion and chronic bronchitis in two rural Bolivian villages", Thorax, 54/11, pp: 1104-1108.

Albalak, R., Keeler, G.J., Frisancho, A.R. and Haber, M.J. (1999). "Assessment of PM10 concentrations from domestic biomass fuel combustion in two rural Bolivian Highland villages", Environmental Science and Technology, 33: 2505-2509.

Araya Asfaw and Yissehak Demissie (2012). "Sustainable Household Energy for Addis Ababa, Ethiopia", Consilience: The Journal of Sustainable Development Vol. 8, Iss. 1 (2012), Pp. 1–11.

Aynalem, A. (2008). Ethiopian Demography and Health. Retrieved (December 2012) at: http://www.ethiodemographyandhealth.org/Ethiopian_ Demography AynalemAdugna.pdf

Bailis, R., Pennise, D., Ezzati, M., Kammen, D. and Kituyi, E. (2004). Impacts of Greenhouse Gas and Particulate Emissions from Woodfuel Production and End-Use in Sub-Saharan Africa.

Beckingham, C.F. and Huntingford, G.W. (1961). The prestes John of the Indies. Cambridge: Cambridge University Press.

Biomass Technology Group (BTG) (2010). Making charcoal production in Sub Sahara Africa sustainable. NL Agency, Netherlands.

Boahene, K. (1998). The challenge of deforestation in tropical Africa: reflections on its principal causes, consequences and solutions. Land Degradation and Development, 9 (3): 247-258.

BoFED (2008). General Profile of the Oromiya National Regional State. Bureau of Finance and Economic Development of the Oromiya Regional State. Retrieved December 2012 at: http://www.pcdp.gov.et/ Oromiya%20 Regional%20Profile.pdf.

Bongers, F. and Tennigkeit, T. (eds.) (2010). Degraded Forests in Eastern Africa: Management and Restoration. The Earthscan Forest Library. Earthscan Ltd. London, UK. 370 pp.

Bouros, D. and Samiou, M.F. (2001). "Short-term effects of wood smoke exposure on the respiratory system among charcoal production workers", Chest, 119/4, 1260-1265.

Brewer, W., Cypher, A., Gress, J., Neirby, S., Petitpren, L. (2010). Investigation of Charcoal Production Methods for Sajalices, Panama.The Mangrove Charcoal Sustainability Engineers for Sajalices, Michigan Technological University.

Bromley, D. W. (1991). Environment and Economy: Property Rights and Public Policy. Oxford: Blackwell Publishers.

Bruce, J. (1813). Travels to discover the source of the Blue Nile in the years 1768-1773. Edinbrurgh. Archibold Constable and Company.

Bruce, N., R. Perez-Pedilla, and Albalak, R. (2000). "Indoor Air Pollution in Developing Countries: A Major Environmental and Public Health Challenge", World Health Organization, 2000. 78(9): p. 1078-1092.

CHF International (2006). Grassroots Conflict Assessment in the Somali Region (Aug. 2006). In: CHF International Annual Report (2006). Building a Better World: Relief to Development, Development in Relief.

Chidumayo, E. and Emmanuel, C. M. (2010). "Dry Forests and Woodlands in Sub-Saharan Africa: Context and Challenges", In: Chidumayo, N and Gumbo, D.J (eds). The Dry Forests and Woodlands of Africa: Managing for Products and Services. London: Earthscan, pp 1-10.

Chidumayo, E.N. (2011). Environmental impacts of charcoal production in tropical ecosystems of the world. Paper presented at Annual conference of the association for tropical biology and conservation and society for conservation biology, June 12-16, Arusha, Tanzania.

Christian Science Monitor (2012). "In Somalia, UN charcoal purchases could be funding Al Shabab terror group". A weekly edition newsletter, prepared by Mike Pflanz, East African correspondent / September 21, 2012; Kenya, Nairobi.

Conservation Strategy of Ethiopia (CSE) (1997). The resources base: its utilisation and planning for sustainability, vol. 1. Environmental Protection Authority and the Ministry of Economic Development and Cooperation, Addis Ababa, Ethiopia.

Costa, J.L, Navarro, A., Neves, J.B., and Martin, M. (2004). Household wood and charcoal smoke increases risk of otitis media in childhood in Maputo, 33(3): 573-8.

CSA (2008). Summary and Statistical Report of the 2007 Population and Housing Census, Federal Democratic Republic of Ethiopia, Population Census Commission (Central Statistics Agency), Addis Ababa.

Czernik, S. (n.d). "Fundamentals of Charcoal Production". IBI Conference on Biochar, Sustainability and Security in a Changing Climate September 8-10, Newcastle, U.K.

Damascene, J. N. (2005). Agroforestry for wood energy production in Rwanda. A Paper Presented in the Workshop on Alternative Sources of Energy in Rwanda, Organized by IRST, 19 – 20 May 2005, CENTRE IWACU, KABUSUNZU, RWANDA.

Daniel Kasahun (2005). Some aspects of the charcoal economy: Its impact on poverty and the environment. FSS paper on poverty, No. 5. Addis Ababa: Forum for Social Studies (FSS).

EFAP (1994). Ethiopian Forestry Action Program. Issues and Actions. Vol. 3. Ministry of Natural Resources Development and Environmental Protection, Addis Ababa, Ethiopia.

El-Juhany, L. I., Aref, M. and Megahed, M. (2001). Properties of charcoal produced from some endemic and exotic acacia species grown in Riyadh, Saudi Arabia.

Ellegård, A. and M. Nordström (2003). Charcoal Potential in Southern Africa: CHAPOSA, in Renewable Energy For Development.   Stockholm: Stockholm Environmental Institute (SEI): P. 4-6

Ellgård, A. (1993/2001). Health Effects of Charcoal Production from Earth Kilns in Chismba Area, Zambia. Charcoal Industry Workshop, revised 2001. Stockholm: Stockholm Environment Institute.

Ellgård, A. (2001). Health Effects of Charcoal Production from Earth Kilns in Chismba Area, Zambia. Charcoal Industry Workshop 1993, Stockholm Environment Institute, revised 2001.

EPA (1998). Background information on Drought and Desertification in Ethiopia. Environmental Protection Authority, EPA, Addis Ababa.

EPA (2008). Ethiopia Environment Outlook; Environment for Development. The Federal Environmental Protection Authority, Addis Ababa, Ethiopia.

Ephrem, H. (2008). Key Energy Issues in Ethiopia: challenges, opportunities and the way forward. In: Tibebwa, H. and Negusu, A. (eds). Agrofuel Development in Ethiopia: Rhetoric, Reality and Recommendations. Forum for Environment, Addis Ababa, Ethiopia. Pp. 1-25.

Ethiopian Rural Energy Development and Promotion Center (EREDPC) (2007). Solar and Wind Energy Resources Assessment (SWERA) Project: Country Background Information, Solar and Wind Energy Utilization and Project Development Scenarios, Final Report, Addis Ababa. Available at: http://swera.unep.net/typo3conf/ext/metadata_tool/archive/download/Ethi opiaSWERAreport_258.pdf.

Ezzati, M. and D. Kammen (2001). Indoor Air Pollution from Biomass Combustion and Acute Respiratory Infections in Kenya: An Exposure Response Study. The Lancet, 2001. 358: p. 619-625.

Ezzati, M. and D. Kammen (2002). "Evaluating Health Benefits of Transitions in Household Energy technology", En. Pol., 2002. 30(10): p. 815-826.

Ezzati, M., D. Kammen, and B. Mbinda (2000). "Comparison of Emissions and Residential Exposure from Traditional and Improved Cookstoves in Kenya". Env. Sci. Tech., 34(4).

Falcão, M. P. (2008). "Charcoal production and use in Mozambique, Malawi, Tanzania, and Zambia: historical overview, present situation and outlook". In: Kwaschik (ed.), proceedings of the "conference on charcoal and communities in Africa", 16 – 18 June, 2008, Maputo, Mozambique.

FAO (1962). Charcoal for domestic and industrial use. FAO, Rome.

FAO (1985). Industrial charcoal making, FAO Forestry Paper No. 63, FAO, Rome.

FAO (1987). Simple technologies for charcoal making. FAO Forestry paper No. 41. Rome.

FAO (1993). Wood energy development: planning, policies, and strategies. Volume 1. Report on the RWEDP Regional Meeting on Rural Energy Planning and Policies. Bangkok.

FAO (2001). Forestry Outlook Study for Ethiopia. Draft Report prepared for the preparation of Forestry Outlook Study for Africa. Unpublished.

FAO (2003). Forestry Outlook Study for Africa–Regional Report: opportunities and challenges towards2020.FAO Forestry Paper No. 141. FAO. Rome.

FAO (2006). Global Forest Resources Assessment 2005. FAO: Rome, Italy. Available from: ftp://ftp.fao.org/docrep/fao/008/A0400E/A0400E00.pdf.

FAO (2008). Industrial Charcoal Production. Development of A Sustainable Charcoal Industry, Zagreb, Croatia.

FAO (2010). State of the World's Forests 2010. FAO, Rome.

FAO Year Book of Forest Products (1950-1974). Rome.

FAOSTAT (2011). Highlights on Wood Charcoal: 2004-2009, Rome: FAO Forestry Department.

Farm Africa (2006). Proceeding of the workshop on Afar Pastoralist. Prosopis Project Emerging Issues EPP Prosopis Project, Awash Ethiopia.

Farm Africa (2008). Experiences on Prosopis management: case of Afar Region.

FDRE (2012). Scaling–Up Renewable Energy Program, Ethiopia Investment Plan (Draft Final). Ministry of Water and Energy, Addis Ababa, Ethiopia.

Federal Democratic Republic of Ethiopia (FDRE) (2011). Ethiopia's Climate-Resilient Green Economy: Green Economy Strategy Report, Addis Ababa, Ethiopia.

Fikre Zerfu, Seid Mohammed, and Abdurehman Eid (eds) (2010). Community Based Rangeland Management: A Manual, Somali Regional State Livestock, Crop and Rural Development Bureau.

Foley, G. (1986). Charcoal Making in Developing Countries. Earthscan Technical Report No. 5. London, International Institute for Environment and Development.

FOSA (2000). Forestry outlook Study for Ethiopia. Draft Report.

Foster, V. (2000). Measuring the impact of energy reform – practical option. in: World Bank. Energy services for the world's poor. Energy and Development Report, p. 34-42. Washington, DC: World Bank.

FRA (2000). Forest Resources Assessment Report. Prepared by FAO.2000. Ministry of Agriculture, Addis Ababa, Ethiopia.

Gathui, T. wa, Mugo F., Ngugi,W. Wanjiru, H. Kamau, S.(2011). The Kenya Charcoal Policy Handbook. Current Regulations for a Sustainable Charcoal Sector, Prepared for PISCES by Practical Action Consulting East Africa.

GEF (2010). Project Identification Form (PIF), Sustainable Charcoal Program. Washington, DC: The Global Environment Facility.

Ghilardi, A. and Steierer, F. (2011). Charcoal production and use: World Country Statistics and Global Trends. Ppaer presented on Symposium: the Role of Charcoal in Climate Change and Poverty Alleviation Initiatives, 5th June 2011, Arusha, Tanzania.

Girard, P., 2002. Charcoal production and use in Africa: What future?, Unasylva, 211, (53), 30-35.

GIZ (2012). Charcoal Production. Energypedia. GIZ HERA Cooking Energy Compendium.

Glaser, B., J. Lehmann and W. Zech (2002). "Ameliorating Physical and Chemical Properties of Highly Weathered Soils in the Tropics with Charcoal: A Review". Biol Fertil Soils, 35: 219-230.

Gufu Oba (1998). Assessment of Indigenous Range Management Knowledge of the Borana Pastoralists of Southern Ethiopia, GTZ, Borana Lowland Pastoral Development Programme.

Haggman, T. (2006). "Pastoral Conflict and Resource Management in Ethiopia's Somali Region". PhD Thesis, Swiss Graduate School of Public Administration (IDHEAP), University of Lausanne.

Harris, W. C. (1844). The Highlands of Aethiopia, London; Longman, brown, Green and Longman.

Hassane, P.N. (2008). Charcoal Supply Chains in Mozambique. In: Kwaschik (ed.), proceedings of the "conference on charcoal and communities in Africa", 16 – 18 June, 2008, Maputo, Mozambique.

Hawando, T. (1997). Desertification in Ethiopian Highlands. RALA report No. 200. Norwegian Church AID, Addis Ababa, Ethiopia.

Hosier, R.H. (1993). Urban energy systems in Tanzania: a tale of three cities. In: Hosier et al. (1993). Urban energy and environment in Africa. Vol.21(5) 1993. pp 510-523.

Hundie , B. (2006). Institutional change in agriculture and natural resources discussion paper explaining changes of property rights among Afar Pastoralists, Ethiopia Icar Discussion Paper 14/2006.

Hussien, A. (2004). Traditional use, management and conservation of useful plants in dryland parts of North Shoa Zone of the Amhara National Region: An Ethnobotanical Approach. M. Sc. Thesis, Addis Ababa University, Addis Ababa, 174.

Ibrahim, A. M. (2003). A brief history of forest service of the Sudan. Khartum: National Forest Corporation.

IIASA, (1995). Global Energy Perspectives to 2050 and Beyond. World Energy Council and IIASA, London.

Jacobs, D. and Kiene, A. (2009). Renewable Energy Policies for Sustainable African Development. World Future Council.

Jama, S. and Walker, R. (1998). United Nations Development Programme, Emergencies Unit for Ethiopia: Changing pastoralism in the Ethiopian Somali National Regional State. Southeast Rangelands project.

Kalumiana, O.S. and Shakachite, O. (2003). "Forestry policy, legislation and woodfuel energy development in Zambia" in: Mugo, F.W. and D. Walubengo (2002), eds. Woodfuel policy and legislation in eastern and southern Africa. Proceedings of a Regional Workshop held at the World Agroforestry Centre, Nairobi, Kenya, March 4-6, 2002. RELMA, ICRAF. Pp22.

Kambewa, P., Mataya, B., Sichinga, K. and Johnson, T. (2007). Charcoal: the reality – A study of charcoal consumption, trade and production in

Malawi. Small and Medium Forestry Enterprise Series No. 21. International Institute for Environment and Development, London, UK.

Kammen D.M. and Lew D.J. (2005). Review of Technologies for the Production and Use of Charcoal. Renewable and Appropriate Energy Laboratory Report. University of California, Berkeley, USA.

Karve, A.D. (2006). "Briquette Charcoal from Sugarcane Trash", Pune, India.

Kiflu Haile, Mats Sandewall and Kaba Urgessa (2009). "Wood Fuel Demand and Sustainability of Supply in South-Western Ethiopia, Case of Jimma Town". Research Journal of Forestry, 3: 29-42.

Kwaschik, R. (ed.). (2008). Proceedings of the "conference on charcoal and communities in Africa", 16 – 18 June, 2008, Maputo, Mozambique.

Logan, W.E.M. (1946). "An Introduction to the Forests of Central and Southern Ethiopia. Type script".in: Magrath, W. (1989). The Challenge of the Commons. Chr. Michelsen Institute.

Lopez, R. (1997). "Evaluating Economy wide Policies in the Presence of Agricultural Environmental Externalities: the case of Ghana", in: M. Munansinghe, et al. (eds.), Greening Economy Policy Reform, Volume II. Washington, D.C.: World Bank

Luoga, E.J., Witkowski, E.T.F. and Balkwill, K. (2000). "Economics of charcoal production in Miombo woodlands of eastern Tanzania: some hidden costs associated with commercialization of the resources". Ecological Economics 35: 243-257.

Madon, G. (2000). "An assessment of tropical dry-land forest management in Africa: what are its lessons? Presented at the World Bank seminar Communication for Village Power 2000, Empowering People and Transforming Markets", Washington, DC: 4-8 December 2000.

Malimbwi, R., Chidumayo, E., Zahabu, E., Kingazi, S., Misana, S., Luoga, E. and Nduwamungu, J. (2010). "Wood fuel", In: Chidumayo, E.N. and Gumbo, D.J. (eds), The dry forests and woodlands of Africa: managing for products and services. London: Earthscan, pp. 155-177.

Mavlankar, D.V., Trivedi, C.R. and Gray, R.H. (1991). "Levels and risk factors for peri-natal mortality in Ahmedabad, India". Bulletin WHO, 69: 435-442.

Melaku B. 1992. Forest History of Ethiopia from early time to 1974. MPhil thesis. University of North Wales, UK.

Melaku Bekele (2003). Forest property rights, the role of the state, and institutional exigency: the Ethiopian case. PhD Dissertation, SLU, Uppsala, Sweden.

Melaku Bekele (2008). "Environmental policies, strategies and programs" in: Taye Assefa (ed.). Digest of Ethiopian policies, strategies and programs, Addis Ababa: Forum for Social Studies (FSS).

Melessaw, S and Hilaw, L. (2011). Final Report: Household Energy Baseline Survey in SNNPR. GIZ: Eco – Bio-energy Department. MEGEN Power Plc., Consultants in Renewable Energy, Energy Efficiency, Climate Change and Sustainable Development, Addis Ababa, Ethiopia.

Merab, B. (1922). Impressions d'Ethiopie. Abyssinie sous Menelik II. Paris: H. Libert.

Million Bekele (2011) Forest Plantations and Woodlots in Ethiopia. African Forest Forum, Working paper series. Vol.1 Issue 12. Nairobi Kenya.

Ministry of Agriculture (MoA) (1986). "የኢትዮጵያ የደንና የዱር አራዊት ታሪክ". (The history of Ethiopian Forest and Wildlife in Amharic) MoA, Addis Ababa, Ethiopia.

Minwuyelet Melesse (2005). "City Expansion, Squatter Settlements and Policy Implications in Addis Ababa: The Case of Kolfe Keranyo Sub-City". Ethiopian Journal of the Social Sciences and Humanities, 2 (2): 50-80.

Mishra, V.K., Retherford R.D. and Smith K.R. 1999. "Biomass cooking fuels and prevalence of TB in India", International Journal of Infectious Diseases. 3(3): 119-129.

MoARD (Federal Ministry of Agriculture and Rural Development) (2005). Woody Biomass Inventory and Strategic Planning Project. A National Strategic Plan for the Biomass Energy Sector. Addis Ababa, Ethiopia.

MoFED (2002). Development and poverty profile of Ethiopia. Household Income Consumption and Expenditure and Welfare Monitoring Survey of 1999/00.

Mohan, M., Sperduto, R.D., Angra, S.K., Milton, R.C., Mathur, R.L., Underwood, B.A., Jaffrey, N., Pandya, C.B., Chhabra, V.K., Vajpayee, R.B., Kalra, V.K. and Sharma, Y.R. (1989). "The India-U.S. case-control study group. India-U.S. case-control study of age related cataracts". Archives of Ophthalmology, 107: 670-676.

Mooney, H.F. (1954). Report on Forestry in Ethiopia, with special reference to the forests of Arussi and Sidamo. Report submitted to the MoA–Imperial Government of Ethiopia.

Msuya, N., Masanja, and Kimangano Temu, E.A. (2011). "Environmental Burden of Charcoal Production and Use in Dar es Salaam, Tanzania." Journal of Environmental Protection, 2, 1364-1369.

Mugo, F. and Ong, C. (2006). Lessons of Eastern Africa's Unsustainable Charcoal Trade. ICRAF Working Paper no. 20. Nairobi, Kenya. World Agroforestry Centre.

Mulugeta, L. and Demel, T. (2004). Natural gum and resin resources: opportunity to integrate production with conservation of biodiversity, control of desertification and adapt to climate change in the drylands of Ethiopia. In: Proceedings of a Workshop on Conservation of Genetic Resources of Non-Timber Forest Products (NTFPs) in Ethiopia, 37–49. Addis Ababa, Ethiopia, 5–6 April.

Mulugeta, L. and Habtemariam, K. (eds) (2011). Opportunities and challenges for sustainable production and marketing of gums and resins in Ethiopia. CIFOR, Bogor, Indonesia.

Naughton-TrevesL, Kammen, D.M. and Chapman, C. (2007). "Burning biodiversity. Wood biomass use commercial and subsistence groups in western Uganda forests", Biological conservation, 134/2, 232-241.

Nawir, A. A., H. Kassa, M. Sandewall, D. Dore, B. Campbell, B. Ohlsson, and M. Bekele (2007). Stimulating smallholder tree planting – lessons from Africa and Asia. Unasylva, 58 (2007/3): 53-59.

Newcombe, K. (1983). An economic justification for rural afforestation. The case of Ethiopia Draft Energy Department Paper, 1-10.

Nketiah, K.S. (2008). The Prospects of Charcoal Production Contributing to Poverty Reduction in Ghana. In: Kwaschik (ed.), proceedings of the "conference on charcoal and communities in Africa", 16 – 18 June, 2008, Maputo, Mozambique.

NL Agency (2010). Making charcoal production in Sub Sahara Africa sustainable. Ministry of Economic Affairs, Agriculture and Innovation.

Nutt , C. P. Wilson, D. C. (2010). Crucial trends and issues in strategic decision making. In: Hand Book of Decision Making. Eds: Paul C.Nutt and David C. Wilson. John Wiley and Sons, Ltd.

Nyssen, J., J. Poesen, J. Moeyersons, J. Deckers, M. Haile and Lang. A. (2004). Human impact on the Environment in the Ethiopian and Eritrean highlands - a state of the art. Earth Science Reviews, 64 (3 4):273-320.

OECD/IEA (2006). World Energy Outlook. Focus on Key Topics: Energy for Cooking in Developing Countries. Retrieved (December 2012) at: http://www.worldenergyoutlook.org/media/weowebsite/2008-1994/WEO2006.pdf.

OECD/IER (2010): World Energy Outlook. Energy Poverty: how to make modern energy access universal? Retrieved (December 2012) at: http://www.worldenergyoutlook.org/media/weowebsite/2010/weo2010_poverty.pdf

Ogundele, A.T., O.S. Eludoyin and O.S. Oladapo (2011). "Assessment of Impacts of Charcoal Production on Soil Properties in the Derived Savanna, Oyo state, Nigeria". Journal of Soil Science and Environmental Management, 2: 142-146.

Oguntunde, P.G., M. Fosu, E.A. Ayodele and N. Giesen (2004). Effects of Charcoal Production on Maize Yield, Chemical Properties and Texture of Soil. Biol. Fertil Soils, 39: 295-299.

Pankhurst, R. (1961). An Introduction to the Economic History of Ethiopia from early times to 1800, London: Sidgwick and Jackson Ltd.

Pennise, D., K.R. Smith, J.P. Kithinji, M.E. Rezende, T.J. Raad, J. Zhang and Fan, C. (2001). "Emissions of Greenhouse Gases and Other Airborne Pollutants from Charcoal-Making in Kenya and Brazil". Journal of Geophysical Research-Atmosphere, 106/D20: p. 24143-24155.

Perez-Padilla, J.R., Regalado, J., Vedal, S., Pare, P., Chapela, R. and Selman, M. (1996). "Exposure to biomass smoke and chronic airway disease in Mexican women". American Journal of Respiratory and Critical Care Medicine, 154: 701-706.

PFE, IIRR and DF (2010). Pastoralism and Land: Land tenure, administration and use in pastoral areas of Ethiopia. Addis Ababa.

Pintos, J., Franco, E. L., Kowalski, L.P., Oliveira, B V. and Curado, M.P. (1998). "Use of wood stoves and risk of cancers of the upper aero-digestive tract: a case-control study". Int J. Epidem. 27: 936 – 940. Abstract available at: http://ije.oxfordjournals.org/cgi/ content/abstract/27 /6/936?ijkey=af141ff541a9d83299ef4f537887b025d300dca9andkeytype2 =tf_ipsecsha

Pohjonen, V. (1988). Establishment of Fuel-wood Plantation in Ethiopia. University of Juensuu, Finland.

Pokharel, R.K. (2011). "Factors influencing the management regime of Nepal's community forestry, Forest Policy and Economics", doi:10.1016/ j.forpol.2011.08.002.

Practical Action (PA) (2010). Promoting Sustainable Charcoal Production and Marketing in Kenya: A Comparative Analysis through Participatory Market Mapping. A report prepared for Policy Innovation Systems for Clean Energy Security (PISCES) by Practical Action Consulting Eastern Africa.

Practical Action (PA) (n.d). Charcoal production. Technical brief, The Schumacher Centre for Technology and Development, UK.

PREDAS (2008). Women and household energy in Sahelian countries – ABP56 special supplement from PREDAS (Program for the Promotion of Household and Alternative Energy sources in the Sahel). As retrieved (on December, 2012) at: http://www.indiaenvironmentportal.org.in/files/ Women%20and%20household%20energy.pdf

Ribot, J.C. (1993). "Forest Policy and Charcoal Production in Senegal". En. Pol., 21(5): p. 559-585.

Rogers, E. and Eliakimu, M. (2008). Woodlands and the charcoal trade: the case of Dare salaam City; working Papers of the Finnish Forest Research Institute, 98, 93–114.

Rural Energy Resources Development Agency of Amhara Regional State (2011). ባህላዊ አክሳዮችን በተሻሻለው የማክሰያ ቴክኖሎጂ ለማላመድና የራሳቸውን ደን እያለሙ እንዲያከስሉ ለማድረግ የተካሄደ ጥናት። የአማራ ብሄራዊ ክልላዊ መንግስት የማዕድን ኢነርጂ ሀብት ልማት ማስፋፊያ ኤጄንሲ: መጋቢት 2003 ፤ ባህር ዳር: ኢትዮጵያ። ("A Study Conducted to Acquaint Traditional Charcoal Makers with Improved Charcoal Making Technology and Help them to Have their Own Forest for Charcoal Making. Amhara National Regional State Mine and Energy Development Promotion Agency. March, 2011. Bahir Dar, Ethiopia, in Amharic)

Russ, D. (1944). "Report on Ethiopian Forests". (Compiled by) Woldmicheal Kelecha, Addis Ababa. Ethiopian Forestry and Wildlife Authority.

Sandford, S. and Habtu, Y. (2000). Emergency Response in Pastoral areas of Ethiopia, DFID (Addis Ababa; August 2000. (IFAD-pastoral Community Development Project-Formulation Report/working Paper 1, May 2002) as quoted by PFE, IIRR and DF (2010).

Sawadogo, A. (2008). Women and household energy in Sahelian countries - A BP56 special supplement from PREDAS. As retrieved at: http://www.hedon.info/BP56:WomenAndHouseholdEnergyInSahelianCou ntries.

Scherr, Sara J., White, Andy and Kaimowitz, David (2004). A New Agenda for Forest Conservation and Poverty Reduction: Making Markets Work for Low-income Producers. Washington, DC: Forest Trends and CIFOR.

Sead Oumer (2007). The 'Privatisation of Somali Region's Rangelands'. In: Ridgewell, A., Getachew, M. and Fiona, F. (eds). Gender and Pastoralism Volume 1: Range Land and Resource Management in Ethiopia. SOS Sahel, Ethiopia. pp. 33–44.

SEI (2008). Household Energy in Developing Countries: A Burning Issue. Policy Brief. SEI Stockholm, Sweden.

Seidel, A. (2008). "Health Aspects of Charcoal Production and Use, on behalf of GTZ". In: Kwaschik (ed.), proceedings of the "conference on charcoal and communities in Africa", 16 – 18 June, 2008, Maputo, Mozambique.

Shiferaw, A., Jeevanandhan Duraisamy, J., Eyerusalem, L.,Yishak, S. and Eyerusalem, M. (2010). Wood Charcoal Supply to Addis Ababa City and its Effect on the Environment. Energy and Environment, 21(6): 601- 609).

Smith, K.R. (1993). "Fuel Combustion, Air Pollution, Exposure and Health: the situation in developing countries", Annual review of energy and environment, 18, 529-566.

Smith, K.R., Samet, J.M., Romieu, I. and Bruce, N. (2000). "Indoor air pollution in developing countries and ALRI in children". Thorax, 6: 518-532.

Sunderland, T. (2011). New research agenda for Africa's dry forests defined at Durban. Available at: http://blog.cifor.org/5614/new-research-agenda-for-africas-dry-forests-defined-at-durban/#.UZobKkrB-uI.

Syampungani, S., Chirwa Paxie W., Akinnifesi F. K., Gudeta Sileshi and Oluyede C. Ajayi (2009). "The Miombo woodlands at the cross roads: Potential threats, sustainable livelihoods, policy gaps and challenges". Natural Resources Forum 33 (2009) 150–159.

Tefera Mengistu (2011). Physiological ecology of the frankincense tree. PhD thesis, Wageningen University, Wageningen, The Netherlands.

Teshale Woldamanual (2011). Dryland resources, livelihoods and institutions: Diversity and dynamics in use and management of gum and resin trees in Ethiopia, PhD thesis, Wegeningen University, Wegeningen.

Theuri, D.K. (2002). "Woodfuel policy and legislation in Kenya" in: Mugo, F.W. and D. Walubengo (eds). 2002. Woodfuel policy and legislation in eastern and southern Africa. Proceedings of a Regional Workshop held at the World Agroforestry Centre, Nairobi, Kenya, March 4-6, 2002. RELMA, ICRAF. Pp9.

Tinsae Bahru, Zemede Asfaw and Sebsebe Demissew (2012). 'Indigenous knowledge on fuel wood (charcoal and/or firewood) plant species used by the local people in and around the semi-arid Awash National Park, Ethiopia". Journal of Ecology and the Natural Environment, 4(5):141-149.

Tomaselli, I. (2007). Forests and energy in developing countries, Forests and Energy Working Paper No. 2, Rome, FAO.

Uhart, E. (1975). Report on charcoal production and Marketing in Ethiopia. Ministry of Agriculture. Addis Ababa.

UN (2010). Sustainable livelihood in the drylands, a discussion paper for the eighth session of the commission of sustainable development. United nations, New York. 25 April–5 May 2000. I FA D.

UNDP (2000). World Energy Assessment. New York, USA, United Nations Development Programme/United Nations Department of Economic and Social Affairs/World Energy Council.

UNDP (2010). Harnessing Carbon Finance to Promote Sustainable Forestry, Agro-Forestry and Bio-Energy CDM Capacity Development in Eastern and Southern Africa.

UNDP/UNDESA/WEC (2000). World Energy Assessment. New York, USA, United Nations Development Programme/United Nations Department of Economic and Social Affairs/World Energy Council.

UN-Energy (2005). The Energy Challenge for Achieving the Millennium Development (also accessible online at http://esa.un.org/un-energy/pdf/UN-ENRG%paper.pdf.).

UNEP [CASCADe. 2009, 26 July 2009]; Available from: http://www.cascade-africa.org/ManageProjects/tabid/91/ctl/Detail/mid/407/ItemID/74/Source/Country/language/fr-FR/Default.aspx

van Beukering, P., G. Kahyarara, E. Massey, S. di Prima, S. Hess, V. Makundi, and K. van der Leeuw (2007). Optimization of the Charcoal Chain in Tanzania. Poverty Reduction and environment Management (PREM) Programme, Amsterdam: Institute for Environmental Studies.

Vira, B. and Kontoleon, A. (2010). "Dependence of the poor on biodiversity: which poor, what biodiversity?" Paper prepared for the CPRC International Conference, Manchester, September 8-10, 2010.

Vivian, H. (1901). Abyssinia. Through the Lion Land to the Courts of the Lion of Judah. London: C. Arthur Pearson Ltd.

Waddams, P. C. (2000). "Better energy services, better energy sectors and links with the poor". in: Energy services for the world's poor. Energy and Development Report 2000, p. 26-32. Washington, DC: World Bank.

WBISPP (Woody Biomass Inventory and Strategic Planning Project) (2004). A Strategic Plan for the Sustainable Development and Conservation and Management of the woody biomass resources, Final report. MoA.

WBISPP (Woody Biomass Inventory and Strategic Planning Project) (2000). (unpublished), Addis Ababa, Ethiopia.

WBISPP, MoA, (2001). SNNPR, A Strategic Plan for the Sustainable Development, Conservation, and Management of the Woody Biomass Resources, Addis Ababa.

WHO (2002). World Health Report: Reducing Risks, Promoting Healthy Life. 2002, Geneva: WHO.

WHO (2006). Fuel for Life (2006), download available at www.who.int

Wilson, R. (1977). "The Vegetation of Central Tigrein relation to its land use", Webia; 32; (1). 247.

Wondwossen, B. (2009). Preparation of charcoal using agricultural wastes. Ethip. J. Educ. and Sc. Vol. 5 (1): 79-93.

Woreta Abera (2007). Amhara National Regional State (ANRS). Bureau of Agriculture and Rural Development. Addis Ababa, Ethiopia.

World Bank (1984). Ethiopia: Issues and Options in Energy Sector. Washington, D.C: World Bank.

World Bank (2000). ESMAP Household Energy Strategy. Leaflet.

World Bank (2009). Environmental Crisis or Sustainable Development Opportunity? Transforming The Charcoal Sector In Tanzania., Washington, D.C: World Bank.

Wylde, A. (1901). Modern Abyssinia. London; Mathew and Company.

Yigard, M. M (2002). Woodfuel policy and legislation in Ethiopia: In: Mugo, F.W. and D. Walubengo (eds). Woodfuel policy and legislation in eastern and southern Africa. Proceedings of a Regional Workshop held at the

World Agroforestry Centre, Nairobi, Kenya, March 4-6, 2002. RELMA, ICRAF. pp 33.

Yirgalem Mahiteme (2008). Visible and Invisible Actors in Urban Management and Emerging Trends of Informalization: A Case Study from Kolfe-Keranio Sub-city, Addis Ababa; NTNU, Norway.

Yisehak Seboka and Duraisamy, J. (2008). Charcoal Supply Chain Study in Ethiopia. In: Kwaschik (ed.), Proceedings of the "conference on charcoal and communities in Africa", 16 – 18 June, 2008, Maputo: Mozambique.

Zerihun W. and Mesfin, T. (1990). "The status of the vegetation in the Lakes Region of the Rift Valley of Ethiopia and the possibilities of its recovery". SINET: Ethiop. J. Sci., 13(2): 97-120.

Zewdu E. and P. Högberg. 2000. Reconstruction of Forest Site History in Ethiopian Highlands Based on 13C Natural Abundance of Soils. Ambio, 29 (2): 83-89.

www.ingramcontent.com/pod-product-compliance
Lightning Source LLC
Chambersburg PA
CBHW080045280326
41935CB00014B/1787